Nonadaptive Selection

CW01391090

Nonadaptive Selection

*An Evolutionary Source of
Ecological Laws*

JOHN DAMUTH AND LEV R. GINZBURG

THE UNIVERSITY OF CHICAGO PRESS   CHICAGO AND LONDON

The University of Chicago Press, Chicago 60637
The University of Chicago Press, Ltd., London
© 2025 by The University of Chicago
All rights reserved. No part of this book may be used or reproduced in any manner
whatsoever without written permission, except in the case of brief quotations in critical
articles and reviews. For more information, contact the University of Chicago Press,
1427 E. 60th St., Chicago, IL 60637.
Published 2025
Printed in the United States of America

34 33 32 31 30 29 28 27 26 25   1 2 3 4 5

ISBN-13: 978-0-226-83856-4 (cloth)
ISBN-13: 978-0-226-83857-1 (paper)
ISBN-13: 978-0-226-83858-8 (e-book)
DOI: https://doi.org/10.7208/chicago/9780226838588.001.0001

Library of Congress Cataloging-in-Publication Data

Names: Damuth, John Douglas, 1952– author. | Ginzburg, Lev R., author.
Title: Nonadaptive selection : an evolutionary source of ecological laws /
    John Damuth and Lev R. Ginzburg.
Other titles: Evolutionary source of ecological laws
Description: Chicago : The University of Chicago Press, 2025. |
    Includes bibliographical references and index.
Identifiers: LCCN 2024034584 | ISBN 9780226838564 (cloth) |
    ISBN 9780226838571 (paperback) | ISBN 9780226838588 (adobe pdf)
Subjects: LCSH: Macroevolution. | Evolution (Biology)
Classification: LCC QH371.5 .D36 2025 | DDC 576.8—dc23/eng20241010
LC record available at https://lccn.loc.gov/2024034584

♾ This paper meets the requirements of ANSI/NISO Z39.48-1992 (Permanence of Paper).

For Susan
and for Lara

Contents

In Memoriam: John Damuth
August 18, 1952–April 2, 2024

John was a brilliant scientist whose wisdom extended beyond academic endeavors to include keen insights into people, culture, history, and the world in general. He was a singularly sharp, clear-thinking, delightful, warm, witty, generous, and wonderfully inquisitive colleague and friend, who died following 15 months of treatment to hold an insidious cancer at bay. John's contributions to evolutionary biology included wide-ranging empirical, theoretical, and conceptual advances in the fields of comparative ecology, ecological allometry, levels of selection, macroevolution, and paleontology.

John's highly recognized discovery of the scaling of population density with body size and its consequence—the independence between population energy-use and body size—was first described in his 1981 article in the journal *Nature* and published when he was in graduate school. He referred to this independence as the energy equivalence of species. By the turn of the century, it was commonly called Damuth's Law, and I used this designation in the allometry chapter of my 2004 book (coauthored with Mark Colyvan), *Ecological Orbits: How Planets Move and Populations Grow*. While working on *Nonadaptive Selection* with John, I suggested to him more than once that we use this designation in our text; given John's characteristic humility, he was reluctant to refer to his discovery as Damuth's Law in a publication of which he was an author. When we spoke for what we both knew would be the last time, in March 2024, I asked again for his permission to refer to energy equivalence as Damuth's Law, and he agreed that I could do so after his death. You will find Damuth's Law described in section 2.3 of this book as one of the prime examples of an ecological law that has evolved by nonadaptive selection.

This dedication would not be complete without acknowledging the invaluable contribution of Dr. Susan Mazer, who is John's wife and an accomplished evolutionary biologist and professor of Plant Ecology and Evolutionary Biology at the University of California, Santa Barbara. In the months following John's death, Susan and I worked closely on the final manuscript—it was then that I became keenly aware how much John's excellent work was supported and influenced by Susan's insights over the years. Her close attention to all of the final stages leading up to the book's publication—including but not limited to creating an excellent index, and providing countless edits and improvements to the text—easily exceeded my own participation. Both John and I owe Susan our greatest appreciation for helping our work to reach the largest and broadest possible readership.

Lev Ginzburg

Preface

This book is presented as an essay, a long argument about the value of a specific type of evolutionary process in biology, which we call *nonadaptive selection*. The idea is simple, yet we think the implications are profound. Nonadaptive selection, as we use the term, is ordinary selection among biological entities, but based on fitness effects intrinsic to the entities under selection, rather than on interactions of traits with a local, shared environment. Thus nonadaptive selection, alone, cannot serve to adapt those entities to a local environment. Instead, it tends to maintain stability and stable equilibria. Nonadaptive selection is of particular value in explaining broad, persistent patterns in multispecies biological units at various levels, where adaptive evolution may be weak or poorly defined. We see it underlying patterns of ecological allometries, community structure, and species interactions, with some implications for macroevolution. Moreover, we find a surprising relationship between these nonadaptive processes and biological laws. We do not advocate the reorientation of any existing research programs, but present nonadaptive selection as an additional conceptual framework that may be useful to add to ecology and evolution.

This book has a somewhat unusual structure. A common type of technical scientific book concerns a specific subdiscipline and takes an "Advances in . . ." approach, updating or reviewing a well-established field in which there is much active research. As mentioned, this book is quite different. It takes as its subject a new (or at least barely explored) idea and argues for the widespread applicability of this idea in the form of an *essay*. The topics that the book touches upon are wide-ranging, and the challenge has been to produce a volume that can be read by as wide a variety of researchers and students as possible. We have kept the text as free from mathematical formulas and jargon as we can; technical details are in a set of appendices for those more familiar with a particular topic or those who want a fuller explanation of it. Thus, we think that one can get a good understanding of our argument from the main text alone. However, we also provide extensive endnotes; these are filled with parenthetical material, but also with significant clarifications and augmentation of the material in the main text. We strongly encourage people to read the endnotes at some stage; they are substantial additions and not just references.

The collaboration that grew into this book began in earnest in 2012, when the authors were both fellows at the Stellenbosch Institute for Advanced Study in Stellenbosch, South Africa. Throughout our careers, we have pursued very different research programs, though we share an interest in theory that addresses general questions. In discussing our work we kept seeing that there was something of a commonality in the way we were thinking about various topics, but it took a while to discover that this common thread was nonadaptive selection.

What follows is our best attempt to present the idea in an accessible form. Since a wide range of subfields is under discussion, it was not possible to give a thorough, detailed review of every topic. The references we cite are meant in many cases merely to lead the reader to a larger literature. We apologize to those readers whose relevant work or ideas were not cited or were treated too briefly.

We also apologize for focusing almost entirely on our own work. This is not because we think the concept of nonadaptive selection applies only to our previous work, or that our work constitutes the best examples of it. However, it is what we know inside and out and turns out to reveal the potential value of nonadaptive selection that we have been working on for so long without knowing it. We sincerely hope that readers, on reflection, will find the concept of nonadaptive selection useful and will uncover many more examples.

CHAPTER I

Selection Processes and Multispecies Systems

1.1 Prologue: Nonadaptive Selection

This book proposes explanations for persistent and recurrent regularities at the macroscopic scale in ecology and evolution. By this we mean causal explanations for recurrent structural patterns observed over large scales of space and time, usually involving biological entities composed of many individuals, populations, or species. For example, food webs are dominated by only a subset of possible configurations; the scaling of population density across species follows an allometric relationship with body mass; and repeated, nonrandom patterns of diversification and extinction can be observed in the fossil record. Much of the traditional focus of both evolution and ecology is on explaining diversification: why organisms over here are different from those over there, or why organisms of the past differ from their modern descendants. In contrast, we think the equally important facts that are critical to explain at macroscopic scales are the similarities among biological systems in space and time: why species interactions, communities, faunas, and so forth take on such similar structural characteristics whenever and wherever we find them. The two tasks are different, so it should come as no surprise that for each the most effective conceptual tools to tackle it may also differ.

As yet there has been little consensus about how best to develop theory that is robust and productive for the macroscopic scale and that supports causal explanation of repeated patterns on that scale. Simple extrapolation of microscale processes that operate locally and on short timescales is usually ineffective in explaining macroscopic structural regularities. Extrapolation fails because familiar, adaptive microscale processes depend on environmental context for their operation, yet environmental change itself is not predictable from within evolutionary theory or its mechanisms. In this book we argue that there is a class of evolutionary explanations that may offer causal accounts of recurrent macroscale patterns, while relying only on processes of selection that are based on discoverable, inherent properties of the entities undergoing selection. We argue that each of these explanatory schemes can be considered a distinct mechanistic description of how part of nature works, based on selection processes that are not sensitive to environmental context. We refer to such explanatory schemes as based on *nonadaptive* selection, and we suspect that this underappreciated conceptual framework can be widely applied in both ecology and evolutionary biology.

A *nonadaptive selection* mechanism, explaining a specific pattern observed in nature, can be defined as a causal explanation that includes (and itself explains) a selection mechanism, where the differences in the probability of persistence or success across members of the population (of any entities) derive only from the consequences of the current internal state or configuration of each population member. In other words, the *fitness of a population member* does not derive from an interaction between that member's phenotype and the local environment. In most cases the internal state or configuration giving rise to fitness differences among entities undergoing nonadaptive selection is some type of inherent stability criterion, which can often be characterized mathematically. For example, nonadaptive selection could be operating in a case in which unstable patterns of species interactions cause local extinction of the involved species, thus causing change in local communities. The communities could be seen to evolve by such means but not in any sense to be "adapting" to a local environment.

Throughout this book we use the term *environment* in the broadest sense, to refer to both the physical and the biological local circumstances encountered by members of the population of entities among which selection is operating. So, for example, if selection operates among the subcomponents of a more complex biological system such as a community, a subcomponent's environment includes not only the conditions external to the community but potentially the current configuration of the com munity itself, which may influence the frequency and types of interactions among the subcomponents.

This is in contrast to the way we ordinarily think of selection in biology, which is to regard it as a process that results in (or maintains) adaptation to local conditions.[1] Adaptive selection occurs because fitness is the result of an interaction between phenotypes and local environmental conditions. In nonadaptive selection, by contrast, there is no "fitness" deriving from an interaction with the environment. Thus the selection process does not (and cannot) result in adaptation. The nonadaptive selection process operates essentially without "seeing" the environment, yet among a "population" of biological systems differing in their inherent stability, the more unstable ones will be preferentially removed and the more stable ones preserved. Furthermore, fully nonadaptive selection has the property of acting the same way in the face of environmental variation in space and time, thus overcoming the problem of extrapolation in explaining macroscopic patterns.

1.2 An Example: Why Food Chains Are Short

A simple example may clarify what we mean. A food chain depicts the sequence of trophic energy transfer among species; it is a graph depicting a pathway of who eats whom, starting from a producer species at the bottom and adding in turn each higher level of consumer species. The length of a chain is the total number of species encountered on a direct passage from the basal producer to the highest consumer. For example, "clover-rabbit-bobcat" is a simple food chain of length 3. It has long been known that the maximum length of food chains tends to be short and shows remarkably little variation among communities, usually exhibiting no more than three to five levels (Pimm and Lawton 1977). Contrary to common naive intuition, food chain length does not appear to be limited by lack of sufficient trophic energy, as one might expect as one ascends the trophic pyramid. Communities of the tropics and of the tundra, which have substantial differences in primary productivity, are dominated by food chains of the same range of lengths (Briand and Cohen 1987; Ulanowicz, Holt, and Barfield 2014).

As we discuss in more detail in section 7.3, food chains that include some degree of *omnivory* (species feeding on more than one trophic level, as is common) become increasingly unstable as chain length increases, such that lengths greater than five or six levels are extremely unlikely to persist. This is a property of the dynamics of the food chain and not a reflection of the level of adaptedness or of the ecological details of the individual species involved. Longer food chains are occasionally observed, so it is possible that they will sometimes form, but they are not likely to last. Communities will end up being dominated by short, relatively stable chains. This reflects an ongoing selective process favoring short chains and eliminating long ones. The difference in probability of persistence ("fitness" in this context) among chains depends only on the properties of the length of a chain, not on an interaction of chain length with the environment or the level of adaptedness of the species involved. We therefore recognize that this selection process is nonadaptive, because it operates independently of environmental circumstances and thus cannot result in local adaptation involving chain length. In fact, it seems that in this case nonadaptive selection dominates the evolution and maintenance of food chain length, because the pattern of short chain dominance to which it leads is more or less universal. Other, *adaptive* selection processes do not appear to override the limits on chain length imposed by selection against the instability of long chains where omnivory is present. This relatively straightforward example exhibits all of the salient properties of nonadaptive selection processes discussed in this book.

1.3 Goals at the Macroscale

The biosphere presents us with a bewildering complexity of life, of biological processes and activities on scales from the microscopic to those that encompass vast regions of the planet. All of this complexity has a history extending billions of years into the past. Some time periods are glimpsed dimly and in mere fragments, while others provide a rich still life of sometimes almost inconceivably great difference to the flora and fauna of the present day. Yet all of the history of life presumably reflects the action of ecological and evolutionary processes that are timeless and universal, in the same sense that the laws of physics are universal or that the action of natural selection operates wherever and whenever conditions are appropriate. It is the goal of the biological sciences to seek and to characterize such processes at all scales of space and time; in doing so as much as possible of the function and structure of biological systems will be brought within the embrace of scientific explanation.

From the perspective of ecology and evolution, we can observe recurrent structural themes above the level of individual organisms, both in the present day and sometimes throughout the geological past. At first this might seem surprising. Organisms are all in a race to extract energy for growth and reproduction, from sunlight and from every type of available source of fixed carbon (including, significantly, each other). Natural selection tends to favor those individuals that increase their share of available resources, without regard to the possible consequences for their own species, other species, community structure, or long-term evolutionary compromise (Van Valen 1980a). Thus the potential for strong ecological interactions, unpredictable dynamics, and rapid, unbounded evolution seems clear, yet when seen from broad geographic or long temporal perspectives, the biosphere does not appear at all chaotic.

Populations fluctuate but seem generally stable in the long run. Multispecies communities in unchanging environments are not constantly collapsing and reshuffling (although this happens sometimes), but rather exhibit a range of quasi-stable structures and (with various degrees of species change) are often recognizable over hundreds, thousands, or even millions of years (Olson 1952; Behrensmeyer et al. 1992; Webb and Opdyke 1995; Brett, Ivany, and Schopf 1996; Jablonski and Sepkoski 1996; McGowran and Li 2000). Moreover, power-law relationships (*allometries*; see appendix 10.1) describe a number of the attributes of organisms as part of biological systems (e.g., the relationships between body size and metabolism, abundance, use of space, and species diversity). These kinds of allometries are not associated with transient states or ontogenetic developmental constraints, but rather represent persistent, near-universal relationships seen across species. To a large extent, they reflect constraint envelopes within which organisms and biotas maintain similar functions or dynamic properties at different scales of body size, space, and time. At macroevolutionary scales we see long-term stability of "adaptive zones" and the ecological roles that species play, as well as repeated or sustained, nonrandom patterns of speciation and extinction. These and other observations suggest that ecology and evolution often operate within some kind of limits, and that there are processes that act to keep most species and ecological systems within "bounds." To be sure, occasional external perturbations (e.g., impacts of extraterrestrial objects, recent human industry, major climate change) force major reorganization on much of the biosphere (Eldredge 2003), but afterward biotas tend to return to the relatively stable structures that we observe most of the time. The individual actors may change, but evidently the rules do not (Wing et al. 1992, 11).

Our ultimate objective, when looking at nature on macroscopic scales, should not be merely to describe or to classify patterns, but to achieve a scientific understanding of their origin and maintenance in terms of causal mechanisms.[2] Explanation of macroscopic biological patterns could be based on adaptive mechanisms that operate on the same large scales of space and time on which the patterns are observed, but it has proven difficult to identify or characterize such processes (if they exist at all for biological systems at this scale). Our argument in this volume is that there is an alternative way to explain many phenomena on these scales that does not require novel, exotic, long-term mechanisms, but rather relies on shorter-term mechanisms that we can potentially observe, test, and often characterize mathematically.

1.4 The Problem of Extrapolation: Why Evolutionary Biology and Ecology Need Selective Processes in Addition to Natural Selection

The theory of Darwinian natural selection, as elaborated by the modern evolutionary synthesis of the early twentieth century (Huxley 1942), is one of the towering intellectual achievements of humankind. This theory explained both the fact of descent with modification and the adaptation of populations and individuals to local conditions—the apparent fit of organisms to the environmental circumstances that they experience. Furthermore, it is generally acknowledged as the only general explanation of adaptation; the theory's reach and depth, its explanatory power, and its flexibility are unusually broad. The importance of the unification of biology thus achieved cannot be overstated.

However, natural selection does not offer satisfying direct explanations of many large-scale patterns in ecology and evolution, such as those described in section 1.2. This is not due to any failure of the theory of natural selection, but because Darwinian selection—on which the Modern Synthesis is based—is directed primarily at issues concerning organismal adaptation to local conditions, and the theory does so with respect to a generational timescale that does not permit extrapolation to much longer time frames. Likewise, conventional ecological processes (e.g., ecological competition, population growth, succession, food web interactions) operate and reach completion on relatively short timescales. Repeated patterns that we see across long timescales (and large spatial scales) are not in general direct extrapolations of these short-term processes. To be sure, the long-term patterns are *consistent* with natural selection and ecological processes on generational scales, but simply knowing this fact does not allow one to explain why some patterns are widespread in space and time and others are rare or absent.

One way to think about this is to imagine that one is, in fact, attempting such an extrapolation of generational-scale evolutionary mechanisms from the present far into the future. The process of microevolution requires that we have available not only the raw material of reproducing organisms and a process that generates variation among them (e.g., genetic mutation), but also a specification of environmental conditions shared by members of the population. Over the long term, this specification must include the relevant ways that the environment might change independently of the evolving organisms. This is necessary because the fitness associated with different phenotypes depends on the interaction of those phenotypes with the environment. However, because long-term environmental change is not constrained by microevolutionary mechanisms or theory and is not readily predictable from other sources, extrapolations of evolutionary change into the far future are at best highly speculative. From an extrapolation perspective, it is thus difficult to generalize about what one expects to see in the long run. Some evolutionary pathways and transitions may be easier or faster than others, but there are few obvious limits to what eventually may be possible. Even armed with considerable knowledge of the starting conditions, it is difficult to avoid the conclusion that in principle, given enough time and suitable circumstances, almost anything that does not violate the laws of physics can evolve (see de Vladar, Santos, and Szathmáry 2017).

Thus, extrapolation of microscale processes is generally an ineffective way to approach explaining structural regularities on long timescales. Something else—other processes that apply to the supraindividual level and operate more or less independently of environmental circumstances—is needed to do that explaining. In this book we argue that there are potentially many such *nonadaptive* selection processes that can provide explanations of particular macroscale patterns. Each such explanatory scheme is based on specific causal mechanisms that operate on relatively short, observable timescales, and those mechanisms form and maintain longstanding selective processes that underlie the ultimate patterns produced or maintained. We argue that evolutionary biology can contain within it many such nonadaptive selection explanations, coexisting with traditional adaptive selection theory, and all combining to produce the shape of nature as we observe it over large scales of time and space.

1.5 The Macroscale

For evolutionary biology, it was explicit in the development of the Modern Synthesis (1935–1947; Huxley 1942) that long-term "macroevolution" was *consistent* with the theory of adaptive natural selection (*microevolution*) but could not be directly explained by it; long-term evolution was (and probably still is) considered by most to be best approached as a form of history (Taylor 1987). This is because the architects of the synthesis recognized that it was not possible to extrapolate directly from the short-term processes of microevolution (operating on a timescale of one or a few generations) to ever longer, and ultimately geological, timescales. True, notably since George G. Simpson (1944), recurrent patterns and themes in the large-scale history of life have been described with increasing sophistication, and analogs with more familiar theories in microevolution and population ecology have been suggested (Simpson 1953; Van Valen 1973; Sepkoski 1978, 1979, 1984; Raup and Sepkoski 1984), but an independent, mechanism-based causal theory of long-term evolutionary patterns has not yet been elaborated or gained wide acceptance. Accordingly, researchers have tended to restrict the scope of the term *macroevolution* to research on patterns of *historical* diversification and extinction of species and higher taxa. This restricted goal is usually intended to provide only historical explanations, not explanations based on mechanistic causal models. We argue that mechanistic explanations of some macroscopic evolutionary patterns are in fact possible, without violating the neo-Darwinian synthesis. We also think that the restriction of *macroevolution* to patterns involving taxa is too narrow, since the problem of timescale and mechanism described previously applies to any long-term evolutionary pattern, including evolution within populations, communities, and other nontaxonomically defined entities, if the timescale is long enough (see Eldredge 1989; Jablonski and Sepkoski 1996).

Furthermore, ecologists increasingly recognize that recurrent and large-scale patterns in ecology pose much the same problems for explanation as do patterns of long-term evolution. A wide range of ecological entities and systems (e.g., ecosystems, communities, food chains and webs) exhibit structural features that are consistent with, but not directly explained by, microscale evolutionary and ecological mechanisms—that is, neither microevolutionary adaptive evolution nor the small-scale and short-term processes of population growth, competition, and so forth. Rather, properties inherent in the ecological systems themselves seem to dominate in forming and maintaining the configurations that we most often see. A growing subfield of ecology that has become known as *macroecology* refers to large-scale, recurrent patterns in the ecological relationships and distributions of species (Brown and Maurer 1989; Brown 1995; Smith et al. 2008; Beck et al. 2012). Like the situation in macroevolution, macroecological patterns are thought to be compatible with but not to be directly inferable from ecological processes involving individual organisms or species. Again, in many of these cases the problem is largely that the short timescale and local spatial scale of the well-known ecological processes do not permit direct extrapolation to macroscopic system properties (see O'Sullivan, Knell, and Rossberg 2019).

So, we have multiple "macro" fields that suffer similar problems of explanation. Also, if one considers change on the macroscale in detail, it is often difficult to distinguish clearly ecological processes of change from evolutionary ones (unless one adopts extreme and arbitrary definitions). Microevolutionary processes are often studied on the assumption that the environment (and thus the covariance of fitness and phenotypes) remains relatively constant (or varies in a known way). Ecological processes are often studied under the assumption that significant evolution will not occur on their timescale. However, it is becoming increasingly clear that even at the microscale, evolutionary and ecological processes are intertwined and often operate on similar timescales (Post and Palkovacs 2009; Moya-Larano, Rowntree, and Woodward 2014; Hendry 2017; Velasco and Pinto-Ledezma 2022). Significantly, for both ecology and evolution, in going to the macroscale there is even more room for both evolutionary and environmental change to occur (and for interactions between the two kinds of changes). Furthermore, much ecological research recognizes that ecological structures themselves evolve (or develop, or assemble) over time, but this represents the evolution of systems themselves and not necessarily the adaptation of organisms within them. Which dominates in a given case—evolutionary or ecological processes—may not be as important as the fact that we need something other than just one or the other microscale theory to explain it.

For this reason, in this book we usually do not distinguish macroevolutionary from macroecological patterns, but instead refer to all as examples of patterns and processes of the "macroscale." The macroscale is basically any scale to which the microscale processes of ecology and organismic selection cannot be directly extrapolated. In particular, it starts at the level in the biological hierarchy of ecological units where species interactions begin to form networks that have specific characteristics of their own (e.g., food web substructure, communities). It also includes higher-level entities such as species and taxa, which are historical constructs rather than ecological associations. Nonadaptive selection processes are not restricted to multispecies entities, but we think they are of particular importance there. The appropriateness of our usage will become clear as we develop our argument and examples throughout this book.

The broad-scale patterns in which we are interested are due neither to microevolutionary adaptation nor to ecological processes straightforwardly extended in time or space. Nor, we argue, can a simple "hierarchical expansion" of adaptive microevolutionary theory, as proposed and explored by many researchers at the end of the twentieth century, form an adequate basis for explaining these kinds of macroscale patterns. The fundamental problem with mechanistically explaining most macroscale patterns is that our known biological and physical mechanisms tend to operate on a short timescale, but we need to be able to translate their actions into results that are observed on a much larger scale of space and time.

1.6 Nonadaptive Selection as a Solution

This book is about one way that we can develop evolutionary and ecological explanations, based on causal mechanisms observable on short timescales, to explain various types of recurrent, relatively stable biological phenomena observed in the long term. We argue that such selection processes are largely independent of circumstances and thus have little or nothing to do with fitting—*adapting*—their target to local conditions. In our view, these selection processes are best described as *nonadaptive* for that reason. The primary feature of these nonadaptive explanations is what one might loosely call *stability selection*—that is, selection involving general properties of specific biological systems that cause variation in their likelihood of developing, persisting, or persisting without major change (Borrelli et al. 2015). The selection process, specific to a particular explanatory scheme and arising from specific causal processes, provides a bridge between the timescale of the causal mechanisms and the biological structure or pattern observable on a long timescale.

In using the term *nonadaptive*, we do not mean to imply *mal*-adaptive (see appendix 10.14). There is no necessity for the results of nonadaptive selection processes to oppose or conflict with the neo-Darwinian processes of adaptation to local conditions. Sometimes they may do so, but the key attribute of explanations based on nonadaptive selection is that the selection regime they postulate cannot reliably fit their target to local conditions because it does not arise from them. Likewise, we emphasize that we also do not mean "neutral" or "random" processes. Nonadaptive selection theories are based on mechanisms that produce long-standing selection regimes that are anything but random.

Our basic view can be summarized as follows. Adaptive causal explanations of macroscale patterns are hindered by the difficulty of extrapolating short-timescale and local processes to larger scales of time and space, as well as the involvement of supraspecific entities for which local adaptation itself is not easily defined. Nonadaptive selection effectively surmounts these difficulties, through a selection regime that derives primarily from fundamental properties of the biological system itself, and is largely independent of external conditions. Significantly, this lack of dependence on local environment means that nonadaptive selection operates quasi-universally throughout the domain of application of a given selection regime, and does so at all times.[3]

Thus, nonadaptive selection can explain repeated or stable patterns across time and space (without "extrapolation"). This leads to a remarkable conclusion: nonadaptive selection processes underlie lawlike explanations for general evolutionary and ecological patterns. Their properties match the concept of a *ceteris paribus* law in the contemporary philosophy of science. Nonadaptive selection mechanisms describe what would happen "all other things being equal," universally within a specifiable domain of application. Furthermore, local and global patterns within the domain coincide. Whether these patterns, generated by nonadaptive selective mechanisms, should be considered true ecological or evolutionary "laws" is a matter of opinion; we think that one could consider these "laws" under a specific, limited concept of scientific laws (see section 3.2), but we don't insist on it. Nonadaptive selection forms a class of *evolutionary* selective processes that can explain the evolution and stable properties of many macroscale biological systems, in *both* ecology and evolutionary biology, and does so across time and space. Given this potential, it seems that these processes may repay closer examination.

Stability selection and related concepts are not new ideas in themselves, but perhaps surprisingly, they seem to be situated mostly outside traditional evolutionary biology. Evolutionary theory as a whole includes well-known evolutionary processes that are nonselective and nonadaptive (such as genetic drift), but the primary focus of evolutionary biology is neo-Darwinian natural selection and its components, which primarily form a theory about adaptation to local conditions. Within evolutionary biology, then, the term *selection* has come to indicate almost exclusively Darwinian natural selection or a related *adaptive* process, even when the term is applied to biological targets other than the individual organism, such as genes, groups, communities, or taxa (e.g., Eldredge 1989). We argue that selection is a more general concept than the particular, adaptive selection that appears in neo-Darwinian natural selection. Selection processes that are nonadaptive in the sense used here do not command very much attention from evolutionary biologists; adaptation is the traditional focus of the field, and adaptation to the local environment (at least in the short term) is almost without exception considered to be the inevitable consequence of sustained selection.

Nevertheless, numerous lines of biological research, particularly in ecology, make reference to the stability characteristics of alternate states and configurations (e.g., May 1973a; Pimm 1982; Shurin et al. 2004; Ulanowicz, Holt, and Barfield 2014). The biological applications of concepts in what we might call the sciences of complexity (e.g., Holland 1992; Kauffman 1993) also seem to allow for nonadaptive selection processes, though the primary focus of that field is elsewhere (see appendix 10.14). However, we think that in general ecologists have resisted identifying selective processes as part of their theories and models. The almost exclusive association in evolutionary biology of selection with adaptation to local conditions may have obscured what seems to us to be the potential for nonadaptive selection to explain observations and phenomena in both ecology and evolution that adaptational theory does not address.

At some level the effect of selection on system stability is self-evident; extremely unstable or unfeasible ecological systems are of course never seen. However, we believe that biologists may have underestimated the degree to which ongoing selection processes continuously trim away those populations or species that have crossed a particular threshold into unstable territory. Such thresholds may be quite close to the conditions in which most species ordinarily exist. For example, it is usually advantageous to individuals to be as fecund as possible, and this leads to the formation of populations capable of rapid growth, but *too* rapid a population growth rate can be destabilizing and lead to extinction (see chapter 5). Note that when we speak of *extinction*, we almost always mean *local extinction*, since these are the events that ecological units experience.

Such nonadaptive elimination processes are potentially powerful evolutionary forces, and they are not specific to biological systems. The evolution of our planetary system has been driven by the elimination (or collision) of many unstable configurations of celestial bodies. The three-body problem is notoriously unstable, which leads to collisions that result either in accretion of bodies or their loss to the system. That is why our current planets are so far from each other, each interacting with the sun but negligibly with the others. What is left is relatively stable (see appendix 10.5 for a fuller discussion). We hypothesize that selective elimination processes in biology are different only in that the forces of biotic change continually regenerate biological systems that approach or transgress eliminative boundaries, and the opportunity for and intensity of elimination processes thus does not diminish over time. Evolutionary theory, as a whole, embraces multiple processes in addition to Darwinian natural selection, and perhaps there is room for some newly described nonadaptive selection processes as well.

In this book we offer a variety of examples of what we regard as explanations based on nonadaptive selection processes. Some seem to us very well characterized in this way, but we provide other examples of a somewhat more tentative or speculative nature. In some cases this is due to the current lack of empirical or theoretical work involving the observed patterns and in other cases to the novelty of applying a nonadaptive selection perspective to a given situation. We are well aware that in some cases there are alternative explanations of phenomena that we attribute here to nonadaptive selection. In other cases the prevailing explanations rely on a well-established line of reasoning that is, in fact, already selective in nature but does not explicitly relate to the conceptual framework that we are advocating. Although the general idea of selection on intrinsic stability properties is not new, we think that our emphasis on the common explanatory features of nonadaptive selection processes and their potential for explanation of large-scale patterns justifies recognition of nonadaptive selection as a distinct type of process. Nonadaptive selection may not be an entirely new idea, but here perhaps we can begin to build a new way of thinking about this old and intuitive idea.

Two Examples: Network Motifs and Damuth's Law

2.1 From Mechanisms to Patterns to Laws

Nonadaptive selection schemes aim to explain how some part of the world works. We think that there may be many such explanatory schemes, and this book presents numerous examples. Each differs in the details of how and why it operates, but we think that there is a general set of properties that all share and that lead to the lawlike character of non-adaptive selection explanations (see section 3.3).

A nonadaptive selection explanation works because (1) a certain set of specified causal mechanisms generates (2) some kind of selective process, which in turn generates or maintains (3) a specific pattern or result. This logical progression does not mean that explanations based on nonadaptive selection necessarily proceed in the order of parts 1, 2, and 3. In fact, historically it seems that almost always the pattern is discovered first, and the selection process and underlying causal mechanism are worked out later.

As we will see, nonadaptive explanations are not particularly exotic. The mechanisms may differ in detail, but they share the three abovementioned components. They do not involve a plethora of novel vocabulary or a catalog of specialized theoretical concepts, nor do the mechanisms conflict with biological intuition. Nonadaptive selection explanations are biological explanations, based on the biological properties of organisms and the systems of which they are a part. Even if we represent the relevant properties by mathematical expressions, our discourse remains within the usual landscape of ecological and evolutionary biology and the way that relevant theory is conventionally used to illuminate biological questions. Indeed, most biologists will likely find much that is familiar in the following pages, though we hope that for many, well-known ideas will be seen to be richer when placed in a broader context. Our overarching purpose is to emphasize a distinction between adaptive and nonadaptive selection and to show how holding to that distinction helps to understand why some kinds of explanations work well for generalizing about macroscale patterns and others don't. A significant implication is that evolutionary and ecological theory at all scales may be largely united by selection but not by adaptation.

2.2 Stability of Modules in Food Webs: Network Motifs

A straightforward example of nonadaptive selection is presented by a pattern observed in community food webs. Food webs describe the trophic interaction network of communities, diagramming the predator-prey relationships among the species. Food webs are a type of complex network that is the focus of a variety of fields, both within and outside of biology, and any such network can be represented by what is known as a directed graph (Caldarelli 2007). Each vertex of the graph (each species) is connected to one or more others by a directional relationship. In a food web, for example, $A \rightarrow B$ indicates that species A eats species B. The effect of species A on the population growth rate of species B is negative, and the effect of species B on A is positive. In food webs, some species could feed on each other, so bidirectional connections ($A \leftrightarrow B$) are possible, where the effects of the species on each other can both be depicted as positive.

Food webs of actual communities are large and complex. However, one can break down a food web into smaller subunits of a few interacting species, called *community modules* by Robert Holt (1997), who discussed some of the more commonly observed types (see also Paine 1980). Such modules (or subnetworks, or subgraphs) form the building blocks of the larger network and can be classified in terms of their connection structure. For example, for predator-prey relationships involving three species, there are 13 possible interaction configurations, illustrated in figure 2.1.

Milo et al. (2002) showed that in large networks, some such modules are much more common than others, giving different networks different

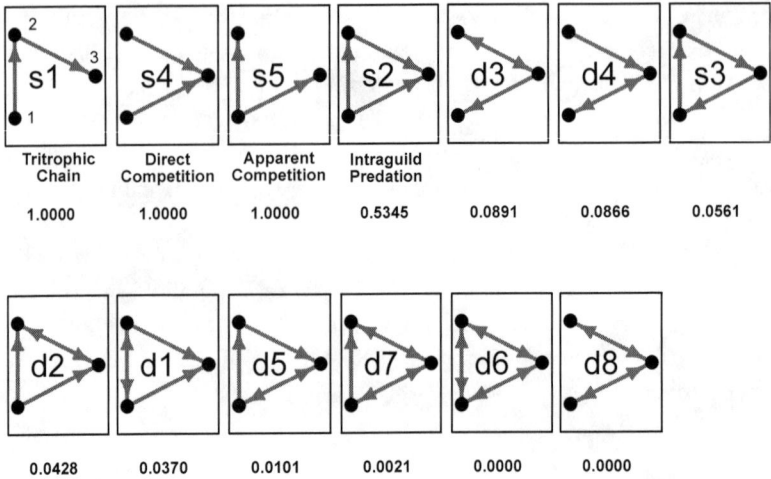

FIGURE 2.1. The 13 possible module types involving three species, after Borrelli (2015) and using his alphanumeric designations. For each module the three species are numbered 1–3 as in s1. Below each triplet module is its QSS as determined by Borrelli (2015), and the triplets are ordered by QSS from left to right and top to bottom. Below the first four triplets are the names by which the respective trophic relationship structures are conventionally known in ecology.

motifs, comprising the set of frequently recurring modules. Significantly, networks of different kinds of things (food webs, the Internet, neuronal connections) tend to exhibit different motifs, because the dynamic properties differ depending upon what is happening along a connection between nodes (Milo et al. 2002, 827). In natural food webs, triplets s1, s4, s5, and (to a lesser and variable extent) s2 are overrepresented relative to the others (Milo et al. 2002; Borrelli 2015).

The simplest explanation for the general overrepresentation of some modules in food webs would be that these triplets are the ones most likely to be produced and to persist—they are most stable. If this is the case, then even if other processes, relationships, or local history also affect the presence of different modules, the selective retention of stable modules and the selective removal of unstable ones may be seen to dominate across communities (note that here, as in many of the cases discussed in this book, selection is among the subcomponents of a given community, not selection of whole communities).

So we want to evaluate and rank the stability of these predator-prey modules in the face of small perturbations. That is, given that we have an equilibrium of species' abundances, will small deviations on the part of any of the species cause the system (the module) to return to equilibrium or to continue to deviate? Note that this does not directly address the question of major perturbations, such as catastrophic loss of a species or ecosystem collapse in the face of major environmental change. But it does include small changes in population dynamics due to chance events or changes due to ordinary adaptive evolution on the part of the species. This *local* (or Lyapunov) stability is fundamental, since if a configuration is not locally stable, it may not last long enough to experience more severe events, and it may be more difficult for it to rediscover a relatively stable equilibrium if it does experience a major perturbation.

The usual way to evaluate such local stability of any food web is to represent the web by a matrix of interaction strengths (Levins 1968; May 1973a; Pimm 1982). In this case, a food web is represented by a matrix whose elements are interaction coefficients between pairs of species. These coefficients represent the influence of the increase in abundance of one species on the growth rate of another.[1]

The reformulation of a food web module to the corresponding interaction matrix is easy to depict. For example, a module that is a three-species trophic chain, in which species 1 eats species 2, and species 2 eats species 3, can be converted to a 3×3 matrix (right), with the interaction terms just represented by their signs. Terms a_{ij} represent the effect of species j on species i, where i and j are represented by the numbers 1, 2, and 3 in the central matrix below:

$$
\begin{matrix} 1 \\ \downarrow \\ 2 \\ \downarrow \\ 3 \end{matrix} \text{ becomes the matrix } \begin{bmatrix} a_{11} & a_{12} & a_{13} \\ a_{21} & a_{22} & a_{23} \\ a_{31} & a_{32} & a_{33} \end{bmatrix} \text{ with signs } \begin{bmatrix} - & + & 0 \\ - & - & + \\ 0 & - & - \end{bmatrix}.
$$

A value of 0 for a_{ij} indicates no direct effect of species j on species i. Where $i = j$ (the diagonal elements of the matrix), a_{ij} is the effect of the species on itself. Assuming some level of density dependence for each species, these values will be negative. (The diagonal elements are usually considered to be similar in strength to the values of the off-diagonal interactions, $i = j$, and that is how we treat them here. Obviously, if the diagonal elements have highly disproportionately large values, they will dominate any interaction's contributions to stability.)

But we don't have the actual values of the interaction strengths; we have only who eats whom—that is, we know only the sign of the interaction coefficient. How then can we tell which modules are more likely to be stable than others?

Sometimes we are lucky. There are some configurations of signs that confer stability, no matter what the a_{ij} values are (as long as the signs are not changed). Such matrices are *qualitatively stable* (or *sign stable*) (See appendix 10.2.) We have more to say about qualitative stability in chapter 7. For now, just note that the first three of the triplets in figure 2.1 (triplets s1, s4, and s5) have this property and will always be locally stable. However, we still know nothing a priori about the local stability of the others, except that it is not guaranteed. We still would like to know how *relatively* unstable they might be.

One way to approach this is to take a signed matrix and assign random numbers to the interactions. If the real part of all of the eigenvalues of this new random matrix is negative, then that particular combination of interactions is locally stable. Do this a large number of times (e.g., 10,000) and count the proportion of these random matrices that are stable. This constitutes a measure called *quasi-sign stability*, or QSS (Allesina and Pascual 2008; see also Pimm 1982; Dambacher et al. 2003). QSS does not directly address the probability that there will be a feasible equilibrium for a module, but it indicates the probability that if an equilibrium exists, it will be locally stable. This in turn suggests how easy it will be to find a new stable configuration of interaction strengths after experiencing a perturbation of larger magnitude.

When Jonathan Borrelli (2015) calculated QSS for the 13 triplets, he found (as expected) that the first three (in figure 2.1, s1, s3, and s5) had a QSS of 1.0. For these three, all of the combinations of randomly generated interaction strengths were locally stable. The other triplets varied in QSS between 0.53 and 0. Borrelli's analysis of 50 published food webs showed a rough correlation between a triplet's QSS and its degree of over- or underrepresentation. This clearly implies that a nonadaptive selection process based on QSS of different module types operates to cause and maintain the patterns of occurrence of those modules in food webs. More recent simulations agree, suggesting that in more complex communities, stability of these motifs is related both to food web stability and to species extinction probabilities (Cirtwell and Wooton 2022).

It is notable that the four most stable triplets have names that are generally recognized in ecology (see figure 2.1, in which they are indicated as trophic chain, direct competition, apparent competition, and intraguild predation; Holt 1977; Polis, Myers, and Holt 1989; Holt and Bonsall 2017;

Cirtwell and Wooton 2022). Researchers would not be expected to introduce names for phenomena whose rarity seldom necessitates reference to them. This suggests that these four triplets are broadly the most persistent and stable in actual communities and thus the more readily observed and studied.

Of course, real food webs are subject to local historical events and additional influences. Interactions among triplets articulating within the larger network might also affect the relative stability of modules of different types, but motif stability still appears to play a substantial role (Borrelli 2015; Cirtwell and Wooton 2022). It is possible that stability differences affect the probability of certain triplets forming to begin with, not merely their persistence after they are established (Maynard, Serván, and Allesina 2018). Although this may be an important distinction for modeling the processes, for nonadaptive selection it does not matter whether the triplets differentially fail or succeed during the assembly phase or later, as long as success and failure in both phases are well represented by the QSS values. This would not constitute a conventional conflict between adaptation and constraint (as suggested by Maynard, Serván, and Allesina 2018); rather the two phases (establishment and persistence) refer to different components of nonadaptive selection. Borrelli's (2015) study also did not address the question of what proportion of his randomly generated matrices was feasible (having all population densities positive at equilibrium); the actual QSS distribution might look somewhat different if restricted to modules with only feasible values. Finally, different biological interpretations of the sign structure, specific to food webs, may be possible (Johnson et al. 2014; Mora et al. 2018). However, what comes across clearly is that no matter how we may in the future refine our understanding of selection on stability of modules, the sign structure of the corresponding matrices alone captures sufficient information to see nonadaptive selection at work.

This example reveals a number of important features of nonadaptive explanations. The overrepresented modules are common not because the fitness of the individual organisms or of the species forming them is somehow optimized; a community food web that comprises a large percentage of these modules is not somehow more "fit" to its environment, nor is a community- or ecosystem-level property maximized.[2] Rather, some modules are more common because there is enormous variation in the stability, and hence persistence, of different modules. This represents a selective process that is nonadaptive, based on inherent properties of each module

Communities

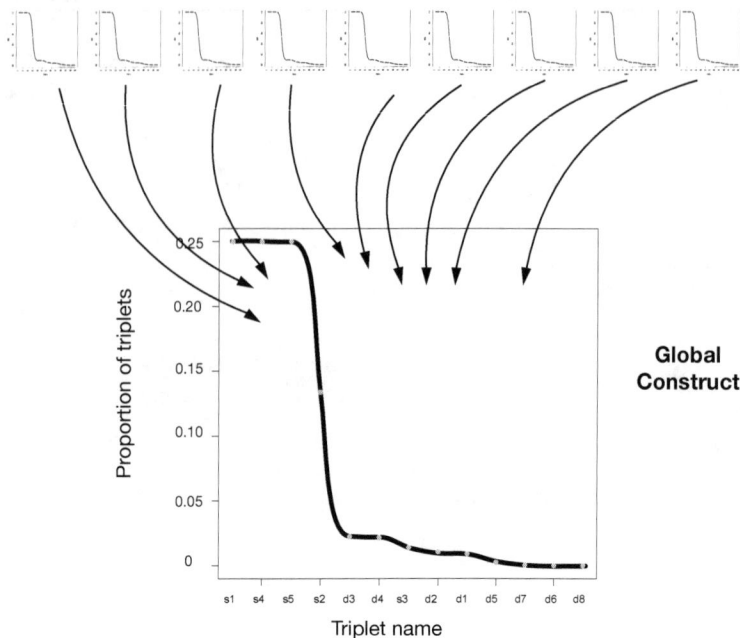

FIGURE 2.2. Sampling triplet distributions globally. Communities everywhere exhibit a similar distribution of triplet types, in which the triplets with the highest QSS dominate. Sampling triplets from this array of communities randomly results in a construct with the same distribution of triplet types as found in the individual communities. The communities need have no spatial relationship to each other, nor are they required to be contemporaneous.

and not on an interaction of the module structure and the environment. Thus it operates the same way in any community, and thereby explains the accumulation and persistence of certain stable modules in food webs in general.

Because the noninteracting species of the whole network, as well as local abiotic environments, appear to have little or no general effect on the stability of such modules, this also means that if we randomly extract species interaction triplets from a wide range of communities and observe the frequencies of the 13 modules, we will also see the same triplets overrepresented; the frequency distribution of the module types should approach the same distribution across communities as within them (figure 2.2). But in the global case we are analyzing a compendium of isolated triplets—we have retained no information about the source of each one. Thus transitioning to the global scale does not obscure the pattern emanating from the local selective processes. Communities are all different from a detailed perspective, but are all "the same" with respect to module dominance: their motifs. Thus location in space is not needed to explain the global relationship, which is happening the same way in each community. The global construct represents fairly the distribution of triplet types within individual ecological communities but is itself not a community.

Note also that because the selection process produces the same result across communities in space, it also does not matter *when* we observe communities, or if we compare communities or triplet distributions in a mixture of coeval and noncontemporaneous sources. Wherever and whenever we look at the module distributions, the distributions will tend to be similar across both space and time. This is the kind of thing that we have been seeking: *an explanation of a macroscale pattern across space and time.*

We risk belaboring this point because of its importance to our argument. This simple example contains all of the key elements of a nonadaptive explanation of a macroscale pattern (in this case, the dominance of only certain triplets in communities). The triplets differ in their stability because of inherent properties of the interaction structure of each triplet. These properties, through a mechanism that is nothing more than their influence on local stability, generate the variation upon which a selection process acts locally among triplets. These properties exert their influence in the same way regardless of environmental conditions or the adaptedness of the species each triplet comprises; differences in triplet stability caused by the inherent properties are not the result of the triplet's interaction with the local environment. This selection cannot result in local adaptation; it is manifestly nonadaptive. The resulting pattern is likely to be the same in the food webs of all communities, in space and time. Global and noncontemporaneous distributions sampling the relative frequency of triplets will mirror the pattern found in local communities. Finally, the correspondence between local stability and triplet occurrence suggests that adaptive evolution, either at the organismic level or at the level of communities (if any), does not significantly overprint the signal left by nonadaptive selection. In uncovering a selective mechanism that cannot lead to local adaptation, we have found a selective mechanism that explains a macroscale pattern across space and time and, at the same time, reveals evidence that no adaptive processes (at any level) significantly overwrite its effects on food web structure.

2.3 Selective Local Extinction and the Allometries of Energy and Abundance: Damuth's Law

Now let us turn our attention to a macroscale pattern involving allometries and at a larger scale. These allometries describe the population density and the energy-use relationships among species of a wide variety of body sizes, within both local ecological communities and more geographically diverse biotas. We argue that these allometries are formed and maintained by a nonadaptive selection process that operates across species within trophic levels, within local communities. This selection process acts in a different way than the one described previously for community modules, by establishing a self-correcting dynamic that maintains an equilibrium. The effects are seen at a global scale and indeed, suggest that this nonadaptive, selection-based process is a necessary condition for the diversity of animal body sizes seen across the biosphere. Yet the proposed mechanism underlying the selection process relies only on well-known properties of species populations and ecological communities.

2.3.1 Body Size, Energy, and Density

We begin by establishing the mechanistic, causal basis and the process of selection that it describes. This allows us to predict the pattern that we expect to see formed and maintained in the long term.

Consider the populations of species that coexist in a local community and that belong to one of the broad trophic levels (say, "herbivores"). At any time these local species populations (which for brevity we just call species) can be characterized by a mean body size and a mean abundance. For simplicity the body size can be expressed as the mean adult body mass of the population. Abundance can be characterized by a population density (the number of individuals per unit area). Each individual of a population must obtain sufficient trophic energy for maintenance, growth, and reproduction, and this is proportional to metabolic rate (assume for the purposes of the example that energy conversion efficiency is roughly the same across all species). The total amount of energy that flows through the population, per unit area, at any time is simply the number of individuals (the population density) times the rate at which individuals use energy (individual metabolism). This is the realized effective rate of energy flow that the population is able to extract from the resource base of its trophic level.

The foregoing are essentially definitions of what it means to be a population in a community. To these we can add two generalizations that will be useful.

First, the allometric scaling of individual metabolic rate (B) across a wide body size range and multiple taxa is relatively well known. An individual of a large species uses energy at a higher rate than does an individual of a small species, but this rate difference is not proportional to the body size difference. In general, among species it is found that metabolic needs scale with species mean body mass with an exponent between 0.50 and 1.00, and a modal value across the largest span of size approximates 0.75 (Hemmingsen 1950, 1960; Kleiber 1975; Savage et al. 2004; Banavar et al. 2010).

Second, if a species evolves to control a greater and greater share of the locally available energy, it becomes progressively more difficult for it to evolve the ability to exploit even more energy without compromising some of the evolving species' existing competitive abilities and giving an opportunity to other, more specialized competitors. That is, it is ordinarily impossible to evolve a supercompetitor, a "master of all trades," that can monopolize all resources in the community and drive most or all other species *to extinction*. This is essentially a process of diminishing returns. For successful species, the necessity to defend the diverse resources of one's large energy base, combined with the decreasing rewards (and difficulty of resolving functional trade-offs) involved in outcompeting an increasing number of other species, suggests that there is a practical upper limit to relative energy use.

2.3.2 *The Primary Mechanism*

Consider the following to be the primary elements of the mechanism underlying an elimination process active in a trophic level in a given ecological community:

1. *The average amount of available energy (fixed carbon) in a habitat (or in the biosphere) is more or less constant over long spans of time* (Van Valen 1976). Corollary: *If one species increases its share of energy in the community (by evolution or by temporary good fortune), one or more other species must lose an equivalent amount from their share.* This can be considered a version of the Red Queen's hypothesis as originally formulated by Leigh Van Valen (1973, 1980a), in which evolution is a zero-sum game for energy. Inclusion of the Red

Queen here ensures that even in a constant environment, over evolutionary time species' energy use will fluctuate as species compete in the struggle to evolve higher energy control and to maintain it.

2. *Species that currently control small amounts of energy relative to other species of their size are at the greatest risk for local extinction.* The risk is elevated both because a small resource base may be more likely to disappear as a result of random long-term environmental changes, and because a small change in the proportional use of resources by a species that controls a relatively large amount of energy in the community exerts a large absolute effect on other species. (Such disproportionately successful energy-controlling species are those that are most abundant for their body size, so it is relatively easy for a small change in behavior or competitive ability of each individual to exert a large summed effect on a small resource base. Such a small resource may be the only energy support for a population of another species.)[3]

This mechanism generates an elimination process in the following way. First let us plot each species' energy use versus its body mass, as shown in figure 2.3. In this example the energy-use values represent long-term average values that the species maintain over substantial spans of time (i.e.,

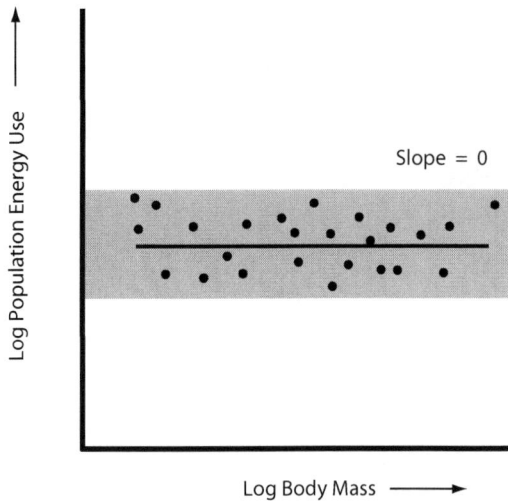

FIGURE 2.3. Energy equivalence among species. Schematic of population energy-use scaling independently of body mass (slope = 0), within a trophic level in a community. Each point represents one species. Shaded area represents the occupied space, for reference.

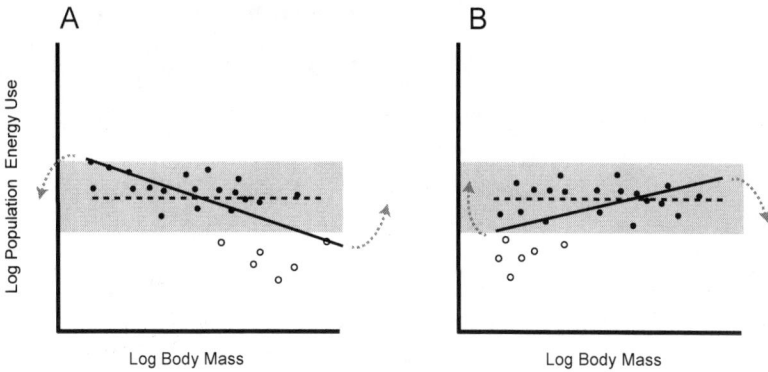

FIGURE 2.4. Effect of chance deviations from energy equivalence. General characteristics as in figure 2.3. A: Some large species (open circles) have by chance become rare for their body size, causing the overall community slope to be negative, and there is now a relationship between energy use and body size. Extinction of these species causes the line to return toward energy equivalence, as suggested by the dotted arrows. B: Same as A, except that now some of the smallest species have become rare.

they are not instantaneously fluctuating values, but ones that will require evolution or major environmental change to alter). The figure represents a single trophic level in a hypothetical community in which there is no trend in population energy use related to body size. This pattern is known in the literature as *energetic equivalence* (Nee et al. 1991) or *energy equivalence*. What it means is that there is no statistical relationship between body mass and energy use; body mass alone does not predict energy use per unit area. All species need *not* be using the *same* amount of energy per unit time; we just stipulate that there is no trend of energy use with respect to body size. Why we have chosen this configuration as a starting point will soon become apparent.

Over time the energy use of species may change as a result of both elements of the mechanism described previously, even if the environment remains constant. As time goes on the community may evolve away from energy equivalence, and a nonzero slope might thus come to characterize the distribution of the community's energy use relative to body size. Or particular historical conditions may have led to the assembly of the community such that it is initially characterized by a substantial nonzero slope.

However, such a configuration is not likely to last. If the slope has become substantially negative (figure 2.4A), it is some of the larger species that currently exhibit the lowest energy-use values. Hence (by the mechanism's element 2, above) these species are now those most likely to become locally extinct if they become the victims of an evolutionary advance by another species or of a random long-term environmental change. Extinction of these species will tend to cause the slope for the remaining fauna to become more positive, reverting to a condition more similar to energy equivalence. Likewise, if the slope is substantially positive (figure 2.4B), a number of small species are currently the ones with the lowest energy use. Their extinction will tend to make the slope more negative (or less positive). When slopes are near zero, species of the lowest and highest energy use are distributed over all body sizes, so extinction of relatively low-energy-controlling species causes only small, random changes in slope. Most of the time, communities will be found in this state, exhibiting only slight deviations, if any, from energy equivalence. The distribution of energy use effectively self-corrects, keeping the slope nearly horizontal. This is not unlike the classic example of negative feedback in cybernetics, the governor of a steam engine. That was exactly the metaphor that Alfred R. Wallace (1858) used in his joint publication with Charles Darwin introducing natural selection. Wallace likened the action of selection in maintaining high adaptedness to such a governor device, specifically in the case of selection for new adaptations to compensate for any "deficiencies" that might have evolved by chance. In modern language, adaptation will tend to return a population to a local adaptive peak, even if for some reason it has been displaced from a peak; adaptive evolution is always acting to increase adaptedness if it can. Our stabilizing nonadaptive selective mechanism is similarly inexorable and effective, and it does not need to wait for favorable mutations to appear.

It is thus the differential local extinction of species populations of relatively low energy use that drives the community to a dynamic equilibrium around energy equivalence and keeps it there, as simulation models suggest (Damuth 2007). This is the selection process that forms the central part of the explanation. Unlike in the case of selection on food web triplets based on their stability, it is not directional in nature but acts as a regulator to maintain a particular invariance among communities in the distribution of energy use across body mass. The result over evolutionary time will be the widespread occurrence of energy equivalence, unless this pattern is overridden or overprinted by other processes.

We think that this simple nonadaptive selection process, generated by fundamental properties of ecological communities and acting over long spans of time, predicts and explains energy equivalence.

2.3.3 Does This Mechanism Apply to Any and All Trophic Levels?

Our expectation of energy equivalence has been derived for a single trophic level, but we have not specified anything about that trophic level. Therefore, reasoning as we did for community modules, any and all trophic levels that exhibit a wide body size range should show the same energy equivalence pattern of zero slope. However, they may well do so at different elevations. Higher trophic levels should have lower absolute, total amounts of energy available to them, and thus the amount of energy controlled by the average species in a higher trophic level should be less; trophic levels should appear to be "stacked" (but with the same zero slope) if plotted together on a graph like that in figure 2.5. In fact, this is true empirically: studies of population density scaling show that higher trophic levels than primary consumers do exhibit similar slopes, consistent with approximate energy equivalence, but at lower densities (Damuth 1987).

2.3.4 Are Local and Global Patterns Similar?

This example has been presented within the context of local community evolution. But just as we saw in the example of nonadaptive selection on food web triplets, sampling across communities should result in a pattern similar to that found in any single community. Such a global construct for a given trophic level would merely reflect the sampling of species populations from a collection of many separate communities, all coexisting on a similar-sized total energy base and experiencing locally the same dynamic elimination process described here. Since this process tends to drive any community, regardless of initial conditions and species composition, to an equilibrium region around energy equivalence, global samples are drawn mostly from communities in rough conformity to energy equivalence and should exhibit the same relationship. To examine this empirically it will be useful to use the scaling of population density as a rough proxy for the scaling of energy-use.

2.3.5 Scaling of Population Density as a Tool to Detect Energy Equivalence

We have a prediction in terms of population energy use (E) and body mass (M): $E \propto M^0$. Yet it is difficult to test this with existing data. Population

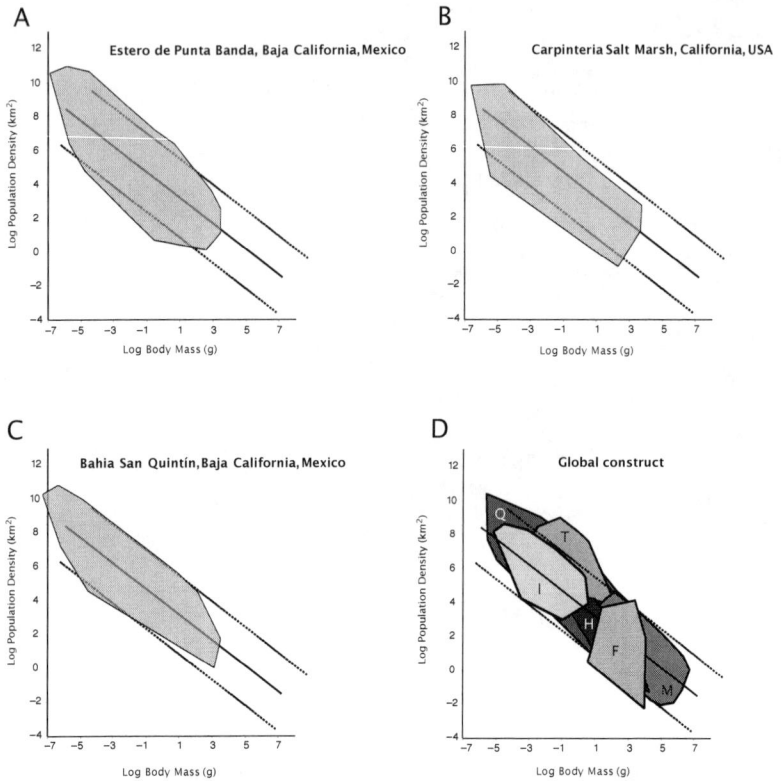

FIGURE 2.5. Polygons surrounding data points for each of three communities (A–C) and a comparison with a large global sample of species (D). All are drawn on the same log10 scale with lines of slope −0.75 drawn as a visual aid. There is a clear tendency in all three salt-marsh communities for the densities to decline as expected, conforming approximately to the −0.75 slopes shown by guide lines in the figure. Hechinger et al. (2011) were also able to make more direct estimates of population productivity and energy use for the species, which revealed that in all three communities energy equivalence holds approximately across body sizes. A–C: Three salt-marsh communities from the Pacific coast of North America (replotted from Hechinger et al. 2011). Polygons enclose population densities of species ranging from small parasites to birds. D: Global dataset sampling 1,397 average ecological population densities worldwide. Species numbers: A, 162; B, 118; C, 130. Numbers, legend, and sources for D: mammals (M), 563 species (Damuth 1993); birds (F), 555 densities (Juanes 1986); reptiles and amphibians (H), 30 species (Damuth 1987); terrestrial invertebrates (I), 62 species (Damuth 1987); aquatic animals (Q), 140 species (Damuth 1987); intertidal animals (T), 47 species (Marquet, Navarrete, and Castilla 1990). See appendix 10.3 for more details.

energy use is onerous to measure directly, and relatively few data are available (cf., Humphreys 1981; Sewall et al. 2013). However, population density estimates are much more widely reported in the literature. We can convert the expected log-log slope of energy use versus body mass (i.e., 0) to a slope for population density (D) versus body mass, since $D = E / B$, and we know that individual energy metabolism (B) scales *approximately* as $M^{0.75}$ over large body size ranges and among diverse and heterogeneous datasets (Hemmingsen 1950, 1960; Savage et al. 2004). The preceding equation becomes the proportionality $M^d \propto M^0 / M^{0.75}$, where d represents the allometric exponent (or log-log slope) of density with mass. Since in division we subtract exponents, $d = -0.75$. In other words, when energy equivalence holds, the exponents of individual metabolism and population density scale with body size approximately to the same exponent as metabolic rate, but with the opposite sign (i.e., the two quantities scale reciprocally). Thus, in communities characterized by energy equivalence, we should expect population density generally to scale approximately as mass to the -0.75 power, although some variation is likely among different datasets simply due to sampling effects, measurement error, and unknown residual error in the estimation of individual energy-use (cf. Sewall et al. 2013). One should note that in the current context, we are interested in how density scaling represents the scaling of population energy use, not in predicting or explaining density per se. Density scaling is an indirect way to investigate energy relationships.

Even using population densities, we still have major limitations on available data. Few datasets include densities of a wide variety of species, spanning a large body size range, within a trophic level of a single community.[4] Probably the most revealing current data come from studies of three salt-marsh communities on the Pacific coast of North America (Hechinger et al. 2011). Data were collected for local species populations in each community with the express purpose of investigating food webs and energy relationships (Kuris et al. 2008). The species span almost 11 orders of magnitude in body size. The sample points in the salt-marsh communities were corrected by Ryan Hechinger et al. (2011) for both trophic level and temperature. These datasets are thus representative of scaling within a single trophic level. (See appendix 10.3 for a more detailed discussion and examples.)

Figure 2.5D represents a global construct formed from sampling a large number of species populations, from a wide range of habitats, throughout the world. Polygons represent the envelopes of density values for various groups. For all except birds, the data represent mean values for each species. The similarity of the global construct and the local communities is striking.

Population density scaling does indeed exhibit a widespread tendency to scale closely to the reciprocal of metabolic scaling and thus supports a widespread tendency toward rough energy equivalence (Damuth 1981b, 1987). Furthermore, the general resemblance between single community patterns and the overall global constructs is what we expect from a non-adaptive selection mechanism. The mechanism we describe has the law-like property of producing a similar result everywhere it operates, and this results in the congruence of global and local patterns.

Energy equivalence is particularly evident when population densities are plotted across several or more orders of magnitude in body mass (White et al. 2007). First described in herbivorous mammals (Damuth 1981b), energy equivalence has subsequently been shown to apply broadly among animals (Juanes 1986; Damuth 1987; Marquet, Navarette, and Castilla 1990; Nee et al. 1991; Cotgreave 1995; Meehan 2006; Meehan et al. 2006; Hechinger et al. 2011; Hatton et al. 2019) and possibly even terrestrial plants (Enquist, Brown, and West 1998). Some exceptions occur in extreme or special environments (Dugan, Hubbard, and Page 1995; Arneberg, Skorping, and Read 1998; Knouft 2002), and selected subsamples of the overall relationship often show the influence of additional, diverse biological processes (e.g., du Toit and Owen-Smith 1989; Damuth 1993; Silva and Downing 1995). Nevertheless, when looking at broad trophic levels, energy equivalence holds very generally and characterizes species of the widest variety of sizes, adaptations, and ecological niches. This underscores that whatever explains the relationship must be highly general and does not depend on the details of species' ecology or of population interactions. The fact that we so often observe the energy equivalence relationship in large, diverse datasets suggests that a nonadaptive process is at work widely in nature, and that the effects of other ecological or evolutionary processes seldom overprint significantly the pattern it produces.

As long as the distribution of energy (or resource density) is independent of size, ordinary processes of population dynamics do not significantly overprint the signal revealed by density (cf. Ginzburg and Colyvan 2004; DeLong and Vasseur 2012; Yeakel, Kempes, and Redner 2018).

2.3.6 Comments on Energy Equivalence

Energy equivalence is itself not a theory, nor is it an assumption. It is an empirical observation—a regularity—that we must attempt to explain. Here we have offered an explanation for energy equivalence in terms of a non-adaptive selection process that operates in the most general of circumstances, without respect to details of the species or their interactions. The same process also explains to a large extent the scaling of population density with body size, because population density and population energy use are related through the scaling of individual metabolism.

Energy equivalence has resisted efforts at explanation in terms of purely microscale ecological processes, since it apparently requires some mechanism that can act to regulate the relative control of energy of different species, regardless of the details of their niches, adaptations, and competitive abilities. Yet known ecological processes such as competition for resources and predator-prey interactions do not include any such regulatory forces; in any case, such processes have no way of balancing their results with the outcome of processes involving other species with which the target species do not interact locally. Moreover, energy equivalence is seen clearly in global constructs that combine data from species of different localities and habitat types, and these species do not plausibly interact in ecological time. Yet the nonadaptive selection explanation outlined here, using only processes and properties observable over relatively short timescales, predicts both the local and global patterns.

Energy is particularly useful in ecological comparisons because (like money in economics) it is fungible (a common currency), whereas the individual organisms of different species and the different trophic resources used by different species (such as apples and oranges) are not. In economics we can compare, say, the net worth of different households in terms of money rather than having to list all the various things each household owns. The rate at which populations of different species extract energy from the available habitat is a good comparative measure of their *adaptedness*, their ability to turn resources into more individuals—as well as a measure of species' potential impact on the environment and upon other species (Odum 1968; Van Valen 1976, 1980a).[5]

In both community and global plots there is a great deal of variation among species (even among those of similar body size) in population energy use, as indicated by population density. Variation of this magnitude has at times caused a misunderstanding that energy equivalence is a claim that all species use *equal* amounts of energy; this is obviously untrue. The variation one sees in the plots is not merely measurement error; much of it is real biological variation. The actual amount of energy that any one species controls depends on innumerable different, uncontrolled factors. The nonadaptive selective mechanism we think underlies energy equivalence is entirely consistent with this large amount of variation: populations are apparently free to wander throughout a generous region of energy space surrounding the line of mean energy use seen in figure 2.4, experiencing no significant deficit, since they do not differ much in extinction probability from their fellows. However, should they wander too far below the mean, their extinction probability will increase (or in the other direction, their ability to co-opt even more energy and continue farther becomes impaired). Using legal terminology, one can view energy equivalence as an "equitable" distribution of energy use rather than an "equal" one, with at any moment each species appearing to hold whatever piece of the pie it was assigned within a body-size independent lottery.

Finally, consider density scaling and evolution from the perspective of energetics and its consequences for the relative success of different-sized organisms. Terrestrial animals span over 13 orders of magnitude in body mass, from tiny soil invertebrates to elephants. Unsurprisingly, individuals of larger animal species require greater absolute amounts of energy per unit time than do individuals of small species. This has immediate consequences for the way that abundance must scale across large ranges of body size. Consider for a moment two species, one a small rodent (body mass 100 g) and the other an elephant (body mass 6,000,000 g). If mice lived at the density of elephants, the elephant population would be extracting energy (per m^2) at a rate almost 4,000 times that of the rodent population (a rough calculation based on the commonly observed scaling exponent of 3/4 for metabolism with body mass; Kleiber 1975). If soil nematodes (mass 1 μg) had the same population density as elephants, the elephant population would be extracting energy at over 20 billion times that of the nematode population. Clearly, in such a world it would pay to be an elephant and not a nematode. But it is equally clear that such disparities in the degree of energy control are unlikely to characterize stable natural systems. Small species would be so feeble, and large species so dominant, that they would be unlikely to coexist for long. Rather, it is obvious that large species must generally have lower population densities than small ones (leading to greater similarity in population energy use among all species), and this agrees with both logic and common experience: there are more mice than elephants in a given area, and more grasshoppers than mice.

Thus, approximate energy equivalence can be thought of as a prerequisite for the coexistence of the wide range of body sizes that we see in the biosphere. All body sizes are "equivalent" in their possible energy use. No species appears to have an advantage (or disadvantage) in gaining access to trophic energy solely as a result of its body size (Damuth 1987). If there were a strong, consistent, long-term deviation from energy equivalence imposed by, say, the nature of resources, the nature of organismic biology, or the physical structure of habitats, one size (or a few sizes) of organism would be strongly favored by both natural selection and stability criteria, and every species would evolve to that size (or be lost). Even a weak but consistent bias in possible energy use would, if given enough time, lead to a decrease in size ranges represented. So the selective, nonadaptive elimination process explaining energy equivalence (see figure 2.4) is in a large part responsible for maintaining not only a widespread allometry, but also, in so doing, the size diversity of life.

And, as we will see in chapter 6, it does more than that.

Nonadaptive Selection: How It Works

3.1 Nonadaptive Selection Is Distinct, Multifaceted, and Already Embedded in Evolutionary Theory

3.1.1 Introduction

The examples in the previous chapters give an idea of what we are calling nonadaptive selection, but it is clear that the details in each case are very different. The explanation of the large-scale patterns focused on different biological processes and on different scales of both space and time. How are we justified in treating these disparate cases as if they represent exemplars of the same conceptual structure? And even if we do admit significant similarities among the cases, why might it be useful to recognize them formally as members of a particular class of explanatory schemes? The answers require not only a fuller discussion of the nature of nonadaptive selective explanations but also a clear differentiation of them from other kinds of explanations of evolutionary change (see appendix 10.14 for a short compendium of what nonadaptive selection might be mistaken for, but is *not*).

Here we briefly consolidate our views on how selection and adaptation are distinct but fit together to allow ordinary Darwinian adaptive natural selection to operate, and this will reveal that *non*adaptive selection is not particularly exotic. Furthermore, we need to discuss the specific properties of macroscale systems in order to see why nonadaptive selection may be of elevated importance in dealing with them. All of this material will be useful as we discuss cases of nonadaptive selection in greater depth.

3.1.2 Adaptation: A Slippery Fish

Our book is not about adaptation, but we need some perspective on the process to understand the circumstances in which we would *not* expect to see it. The word as now used in evolutionary biology was introduced by Wallace in 1858 (Kutschera 2003). Adaptation is frequently described as a "slippery" concept, by both biologists and philosophers alike (Reeve and Sherman 1993; Rose and Lauder 1996; Hull and Ruse 1998).[1] As a technical term it generally retains the ordinary sense of the word as a process of "fitting one thing to another" (Amundson 1996). In general vernacular use *adaptation* can refer to any change in response to an environmental stimulus, but in evolutionary biology it refers to a specific kind of evolved response resulting from natural selection. This evolved response, resulting from natural selection, is the only way that we use the term.

The fact that there are these two terms, selection and adaptation, and the process referred to by one (selection) is necessary for the process referred to by the other (adaptation), suggests an intimate connection between the two distinct concepts. To be sure, when the fitness differences associated with a given trait result from a specific, causal interaction of phenotypes with local, historically contingent conditions, it seems that an adaptive response to selection response must be occurring with respect to at least some traits. It is difficult to avoid this conclusion with respect to selection among conspecific, co-occurring organisms in the ecological population that they share; adaptation there is pervasive, and organisms are often clearly fitted well to local conditions experienced by their population, now and in the past. Any nonadaptive selective processes that may be occurring at the organismic level are thus likely to be overwhelmed by diversifying adaptive evolution, and the effects of nonadaptive selection alone will be difficult to observe. But with respect to biological systems in which selective processes do not easily or cannot result in local adaptation, we would expect to see the effects of nonadaptive selection (if present) forming repeated patterns involving the target systems across space and time, rather than diversification of those systems.

A number of themes recur throughout the literature of adaptation. Taken together these themes characterize the essence of what most scholars think about adaptation as a process. When any or all of the following five major themes are not applicable to a process of selection-based change, we have reason to doubt that adaptation is or can be occurring as a result of that selection. So let us consider what seem to be the salient features of adaptation.

1. ADAPTATION IS LOCAL. We follow Dobzhansky (1968, 111) and most biologists in recognizing that it makes no sense to think of something as being adapted in the abstract; adaptation is always *to* a specific environment or context (see also Sober 1984). The definitions of adaptation given in the literature invariably make reference to an environmental component or context (Lewontin 1978; Sober 1984; Brandon 1990; West-Eberhard 1992). Thus, for most researchers adaptation is always a local process and results in a fit to some local environment.[2] That environment may need to be defined broadly, but it is still specific to time and place, and thus adaptation is a process that produces evolutionary modifications that are themselves historically contingent. That is, an adaptive evolutionary change is specific to and originates in a specific place and time, even though it may later become more widespread. All adaptation is local. Adaptations *originate* in specific environments in space and time as novel traits are formed by natural selection on variant trait states; they proliferate and are maintained and further *elaborated* by ongoing selection on new variants that contribute to fitness in a particular context. Therefore, when we speak of adaptation we mean a process that does all of this. This scheme is, in fact, ordinary natural selection as described by adaptive theory. Adaptation can occur because the fitness differences are caused by an interaction with the environmental context in which the adaptation provides the solution to a problem or a functional advantage. Adaptation occurs only context by context and is as historically contingent—and as un-extrapolable—as microevolution itself.

2. ADAPTATION ≠ SELECTION. Selection and adaptation are distinct processes, although the nature of the distinction is not always clear from the way the terms are used. Identifying adaptation requires indicating a specific, causal relationship between phenotypes and the environment and is not an integral part of the selection analysis per se (see Lewontin 1978, 1980; Brandon 1990). However, in the recent past, at least, it has been typical for evolutionary biologists to use the terms "relative fitness" and "relative adaptedness" almost interchangeably (Brandon 1978).[3]

Fitness (or a fitness function) represents relative values expressed or estimated in a specific unit. That is, fitness differences are fungible, in the sense that they represent a value in some general currency (such as reproductive success, survival, the ability to find mates, etc.); estimated fitnesses can be averaged (e.g., across years, relatives, or environmental conditions), and fitness components can be combined, because they are in the same units. Adaptations, on the other hand, are traits whose influence on fitness derives from the actual biological problem or challenge that ultimately engenders the fitness values themselves. *Adaptedness*, or the current state of the fit of an organism's traits to its environment, can be regarded as an absolute measure of some performance criterion that is specific to the current or historical context (Brandon 1978, 1990; Reeve and Sherman 1993). Adaptations themselves, though, are not comparable, fungible units of value.

3. ENVIRONMENTALLY IMPOSED PROBLEM. Adaptation adds to selection mechanics the idea of a process that ameliorates or solves a "problem" posed for the organism by a particular environment (one could express it as a "task" that the environmental context obliges the organism to perform or a "benefit" for performing the task relatively well; e.g., Lewontin 1978, 1980; Sober 1984; Burian 1992; West-Eberhard 1992). Of course, these are not necessarily problems or benefits perceived by the organism as such, but rather our assessment of the challenges the organism faces in such a context and how selection either currently favors or in the past has favored certain traits that have the property of successfully addressing those specific challenges. Thus, we can recognize that the current level of absolute adaptedness or fit to the local environment was likely accomplished by a history of selection (not necessarily very long) in that environmental context for particular traits. In adaptation a "problem" may not be evident until a solution (a variant conferring higher fitness) presents itself.

4. NOVELTY. The ongoing process of adaptation produces traits that are *adaptations*, meaning that they are the traits that we recognize as contributing to the close fit of organisms to the current or previous conditions of existence in their local environments. Widespread adaptation can straightforwardly lead to diversification as well, as environments themselves change and diversify. An adaptation that arises in one environment may be useful and thus retained or further elaborated in a wider range of environments. But such an adaptation nevertheless arose at a particular time and in a particular place; it is still the product of a historical process.

Adaptation includes the idea that the process occurs over time, perhaps throughout a succession of selection events or generations, culminating in molding or forming a novel character state or trait that is a distinctly identifiable "adaptation." Thus the historical component of adaptation is a transgenerational process of evolution that requires heritability and the generation of novel variants, whereas selection, strictly speaking, does not. A full theory of adaptation must thus include the phenomenon of producing historically novel traits; otherwise, there would be no way for adaptation to follow or be maintained in the face of changing environmental conditions and contexts.

5. INDEFINITE HISTORICAL MODIFICATION. In principle, adaptive evolution can cause mean trait values to depart indefinitely from the ancestral state, and adaptations can be built up over a long history of the process (Wallace 1858). Thus not only do adaptations require the whole of natural selection (not just the selection process) in order to be formed, but adaptations can be built upon adaptations, and traits can change their function and adaptive status over time. Adaptations are historically contingent evolutionary changes occurring in local environments; just like neo-Darwinian microevolution in general, the course of adaptive changes cannot be extrapolated to long time spans, since the process requires a (possibly changing) environment, and may involve any kinds of traits. It may be the population that adapts, but in the process the dominant organisms change, and under continuous adaptation, organismal variants continue to be changed indefinitely.

The same process of adaptive natural selection both causes the proliferation of novelties throughout a long history of adaptive evolution and maintains those traits as relative advantages within extant populations. At the same time, natural selection does not always optimize fitness, nor does it necessarily create a phenotypic novelty. Yet it operates in a situation in which if the environmental context were to be maintained and sufficient variation produced, the process would be capable of producing novel adaptations. Our aim in this book is not to identify the most useful operational definition of adaptation, but rather to argue that under any reasonable definition, there nevertheless may be at the same time intrinsic properties of biological systems that may generate additional selection regimes that cannot lead to novel adaptations and hence must be nonadaptive.

In sum, one could argue that the preservation of useful novelties at the local scale is the essence of adaptation. For organisms, this capability is provided primarily through aspects of the process of reproduction. However, this fact does not preclude the existence of other mechanisms that would accomplish the same thing, and it may be associated with biological systems at levels other than populations of individual organisms, such as multispecies and macroscale systems (see, e.g., Bourrat 2014; Papale 2021). As we see later in this chapter, though, there is little agreement about heritability and adaptive processes at these higher levels.

3.1.3 How Selection and Adaptation Articulate in Evolution

Let us begin with how biologists usually describe the theory of evolution by Darwinian natural selection. The theory is often presented in terms of a set of three principles,[4] which are intended to be faithful to Darwin's reasoning: phenotypic variation, differential fitness, and heritability. A widely cited formulation is that of Richard Lewontin (1970):

1. Different individuals in a population have different morphologies, physiologies, and behaviors (phenotypic variation).
2. Different phenotypes have different rates of survival and reproduction in a specific environment (differential fitness).
3. There is a correlation between parents and their offspring in the contribution of each to future generations (fitness is heritable).

These three conditions must hold for there to be *evolution by natural selection*. They do not, however, directly address *adaptation*, which Lewontin (1978, 1980) describes as a fourth principle,[5] related to Darwin's *struggle for existence*. This fourth principle involves the *causes* of the fitness differences in criterion 2 and their implications for evolutionary change. The relationship of adaptation to the mechanics of evolution (conditions 1–3) has been controversial and difficult for both biologists and philosophers to express. For the moment we can summarize the primary linkage between selection and adaptation thus: the first three principles explain *how* evolution by selection works, but adaptation explains specifically *why* that evolution leads to a fit between organisms and their local environment (see Brandon 1978, 1990 for a more detailed exposition of this point).

Figure 3.1 shows schematically the way that the Lewontin conditions can be understood to represent the distinction we make between adaptive and nonadaptive selection. Ignoring heritability for the moment, figure 3.1A shows what the first two Lewontin conditions require for selection to be effective: phenotypes have fitnesses. But the source of those fitnesses need not be specified. Fitnesses are fitnesses.[6] They do not need to come from anywhere in particular. If the different phenotypes simply

A   Phenotypes $\xrightarrow{\;\;n\;\;}$ **Fitnesses**   *Nonadaptive selection* (n) alone,
no interaction with environment

B   Phenotypes $\dashrightarrow{\;\;n\;\;}$ **Fitnesses**   Typical ***adaptive natural selection***,
n weak or absent

Local Environment

C   Phenotypes $\xrightarrow{\;\;n\;\;}$ **Fitnesses**   (Proposed) selection among
multispecies entities,
adaptive evolution tends to
be weak or absent

Local Environment

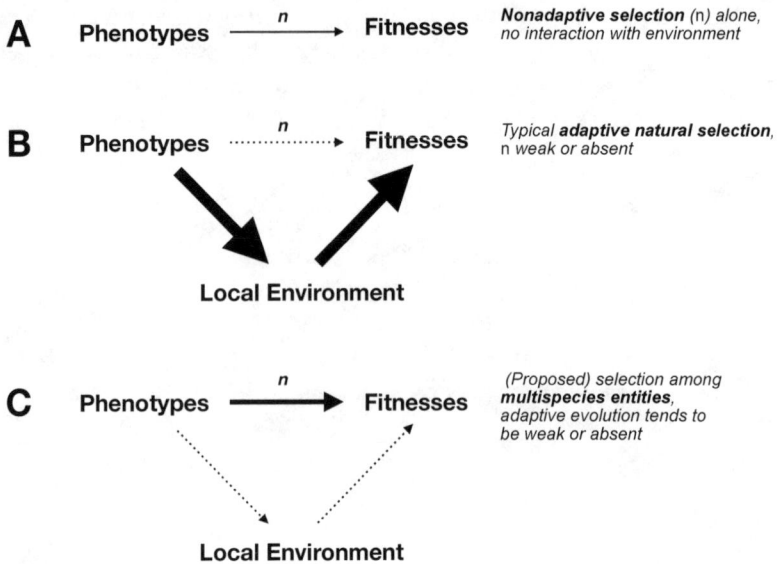

FIGURE 3.1. First two Lewontin conditions and adaptive and nonadaptive selection components. The line marked *n* indicates the nonadaptive selection component. A: Nonadaptive
selection alone; variation in fitness depends solely on intrinsic, configurational properties of
the phenotype. B: Ordinary, Darwinian adaptive evolution; most variation in fitness associated with a given phenotype derives from interaction between the phenotype and the local
environment. C: Evolution by selection suggested for macroscale entities such as multispecies
interaction networks, where adaptive evolution may be weak or absent.

have fitness differences as a condition of their existence, based on inherent properties that they possess (capacities or stability properties that they
were "born with," so to speak, marked *n* in the figure), then *selection* (in the
sense of working through the implications of the variation in fitness) can
occur. However, in this simple case (figure 3.1A), adaptation to the local environment cannot. The environment is not involved. What we usually mean
in evolutionary biology (and what Lewontin discussed at length, because he
was interested in defining selection that *could* be the basis for adaptation) is
depicted in figure 3.1B. This is now selection that can form the basis for Darwinian adaptive natural selection. There is an interaction between the phenotypes and the local environment, symbolized by the heavy arrows, that
provides the connection to the local environment, so that adaptation can
occur. Note that we mean "environment" in the broadest sense: the entire
selective environment, including physical factors; other biological systems;
and even the effects of, say, an organism's own actions.[7]

Note that nonadaptive selection (figure 3.1A) does not require that the members of the population experience a shared environment, as in figure 3.1B. In nonadaptive selection, there is no interaction with the environment. This means that evolution by nonadaptive selection is independent of environmental variation in space and time; it works the same way regardless of the environment.

Considering populations of organisms, we ordinarily think of the effect on fitness of inherent properties of the phenotypes (the dotted arrow marked *n* in figure 3.1B) as being negligible, or at least as easily overwritten or co-opted by the effects of the powerful and pervasive process of adaptation. That is, nonadaptive components of selection are usually of little interest even should they be present. However, what if the adaptive components of selection are weak or absent, as in figure 3.1C? We argue that this is likely to be the case for macroscale systems involving species interactions, in which it is progressively more difficult to apply the concept of local adaptation at the appropriate level. At the macroscale, we might predict that we would see predominantly the effects of nonadaptive selection, occurring among macroscale systems or their components. Even should there be legitimate adaptive processes at higher levels, their effects are evidently weak, as many examples in this book suggest. We do not argue that there *cannot* be any adaptive evolution involving macroscale systems at their level, only that the effects of such selection appear to be small relative to nonadaptive selection.

Notably, Lewontin's three conditions do not specify biological details of the entities participating in the process of evolution. No specifics are offered for the source of phenotypic variation. Likewise, no details of heredity or reproduction are specified; some kind of "like gives rise to like" is all that is necessary for selection to operate. This has led to the general recognition that evolution by natural selection involving "units of selection" other than individual organisms is logically possible. Selection (and evolutionary change) at different levels may present no logical difficulties, but this does not necessarily mean that adaptation occurs at all levels or has the same importance at all levels.

3.1.4 Selection: A General Process

Selection is a key component of evolution by Darwinian natural selection and is the part that gives the whole theory its common name. For many people, *natural selection* is synonymous with the entire process of adaptive evolution, including all three of the Lewontin conditions and numerous additional processes, including, significantly, adaptation. Natural selection in this broader sense includes not only the selection process per se, but also the generation of variability, the transgenerational processes of inheritance, the struggle for existence, and adaptation to local conditions. Darwin himself defined "natural selection" this way (1859, 127), and many biologists maintain this usage (e.g., Endler 1986).

In contrast, contemporary theory increasingly associates the process of selection per se with a general representation known as the Price equation (described in detail in appendix 10.4) (Price 1970, 1972; Wade 1985; Frank 1995). This usage corresponds more closely to the second Lewontin condition alone (Lande and Arnold 1983; Arnold and Wade 1984; Mitchell-Olds and Shaw 1987; Kingsolver et al. 2001; Frank 2012).

In its simplest form the Price equation deals only with the change across a generation in the distribution of phenotypes, as a result of selection— that is, it describes the dynamics of change based on the fitness values borne by different phenotypes (Price 1970). (This is often called the "covariance" formulation of selection.) The full version of the Price equation (Price 1972) adds an additional term to represent the effects of other processes of change that modify or interfere with the outcome of selection per se (such as partial heritability). This full equation provides a complete description of change from one generation to the next, yet the selection still appears as a distinct term. This is logical; for example, selection can occur within a generation even if there is no heritability (Fairbairn and Reeve 2001), and likewise, phenotypic variation can be heritable even if it does not result in fitness differences. One advantage of the Price equation is that it does not specify the biological properties of the traits being selected, which may be gene frequencies or any kind of complex phenotypes that can be enumerated, in any kind of population (Frank 1995). Likewise, it is straightforward to derive many different and more elaborate, specific models of the selection process from the Price equation, to address or highlight particular evolutionary questions. David Queller (2017) argued that the Price equation is *the* fundamental theorem of evolution. The equation describes the dynamics of selection processes involving any entities, including nonbiological ones (as Price [1995] himself argued). In this view, *selection* constitutes but one (critically important) component of overall evolutionary change, and this is what we are talking about in this book as *selection*.[8] We accordingly restrict the terms *Darwinian*, *natural selection*, and *Darwinian natural selection* to refer to Darwin's and others' more inclusive usage, or, equivalently, the whole process underlying adaptive evolution among organisms. To be sure, selection as we use it is still entirely a natural process, but by the specific combination "natural selection" we mean *the adaptive evolutionary theory based on accumulating change by selection among variants of different fitness that explains the adaptation of individual organisms to their local environments.* In contrast, when we use the unqualified term *selection*, it is in the general sense to mean the selective component of either adaptive or nonadaptive evolutionary processes, suitably qualified as necessary, and corresponding to "selection" as represented in the Price equation (see appendix 10.4).

Selection is based on a net fitness that may nevertheless be the combination of multiple fitness components and causal influences. In practice, we often describe selective contributions to overall fitness as if they were separate processes of selection, though they may be happening simultaneously. This is particularly the case if the components are of different major types—for example, viability selection, sexual selection, or group selection. Clearly, in this sense there can be multiple selective processes operating simultaneously in a given biological system, even at a single level of organization. Different traits can contribute independently to overall fitness; traits can contribute to fitness differences because of a causal association with different factors of the environment. And a given trait's contribution to fitness may depend on the simultaneous expression of other traits. (Further, there can be nonselective influences on overall fitness— e.g., environmental effects—that also affect the evolutionary change that is realized in the next generation.) We argue that nonadaptive selection processes have distinct properties and roles to play in evolutionary explanation that justify their being distinguished as a specific category of selection processes, but in nature nonadaptive selection can be commingled with other processes of change and adaptive selection of various kinds.[9]

A further point of significance is that an analysis of selection dynamics—no matter how modeled—does not represent a fully articulated model of *why* the selection is taking place. Only a causal model *outside* of the representation of selection dynamics can reveal this. Observing a selection process at work does not necessarily indicate that a local adaptive process is also occurring. See appendix 10.6 for an example from multilevel selection that shows that one cannot use selection dynamics alone as an "adaptation detector."

Selection is a kind of process and is not restricted to its appearance in Darwinian adaptive natural selection, or even to biology. Selection applies more generally to any collection of entities subject to a selection process, whether or not those entities are thereby also undergoing a process of adaptive change that progressively fits them to the conditions of their local environment. Generalizing selection in this way is in harmony with selection as widely understood in biology, and also with how it appears in areas outside of biology and throughout the history of Western thought. (See appendix 10.7 for the scholarly history of selection without adaptation and appendix 10.8 for discussion of selection outside of biology.)

3.1.5 Selection versus Transformation

Not all change in biological systems is due to selection, of course. Joseph Fracchia and Richard Lewontin (1999; reprinted in Lewontin and Levins 2007, 267–96) described "evolutionary" theories in general to be ones where there is "change in the nature of ensembles of elements." This broad definition of "evolution" is adequate for present purposes, and it emphasizes that what evolves is some kind of system with, possibly, interacting parts. But all change is not the same. Fracchia and Lewontin wanted to distinguish theories of change based on transforming a system from those theories whose changes are caused by selection processes. They introduced a useful distinction, classifying evolutionary theories as being either *transformational* or *variational.*

A transformational theory is one in which change in the system or ensemble occurs because each of the elements changes in roughly the same way or with respect to some kind of regular developmental sequence or program. Transformational theories describe an unfolding of a chain of events inherent in the directed or spontaneous changes in the system's elements. No specific response to environmental conditions is implied, although transformations can be set in motion (or blocked) by external events. As a nonbiological example, Fracchia and Lewontin (1999) suggest that the life cycles of stars exemplify a transformational process; depending on their initial mass, stars pass through a series of stages as they form, change size, and ultimately collapse. The components may change in form or relationship to each other, but the overall sequence is "an unfolding or unrolling of a history that is already immanent in the object." In biology, ontogenetic trajectories and short-term physiological responses are clearly transformational and not selective. Since adaptation to local conditions, in the sense of our discussion, requires selection, such transformative "evolution" cannot cause adaptation in the sense that evolutionary biology uses the term.[10]

The alternative to transformative theories is variational theories, which are based on selection and can (but do not necessarily) support the possibility of adaptation. In variational theories evolution is the result of a selective process among system components. Ensembles are composed of elements that have different properties. As Fracchia and Lewontin (1999, 61) argued:

> The evolution of the ensemble occurs because the different individual elements are eliminated from the ensemble or increase their numbers in the population at different rates. Thus, the statistical distribution of properties changes as some types become more common and others die out. Individual elements may indeed change during their lifetime, but if they do, these changes are in directions unrelated to the dynamic of the collection as a whole and on a time scale much shorter than the evolutionary history of the group. So, the developmental changes that characterize the aging of every living organism are not mirrored in the evolution of the species.

This perspective is consistent with our view of the generality of selection and applies easily to biological and nonbiological cases. To stay with astronomical examples, we see the formation of different planetary systems around a star as a result of nonadaptive selection. The planets that form, their sizes, and their locations are not predetermined at the beginning of the star's history, but form as a process of selective elimination through collision and amalgamation of many bodies in the protoplanetary disc. This process reaches an equilibrium when material is concentrated in large objects that have negligible gravitational interactions with each other. (See appendices 10.5, 10.7, and 10.8 for a more detailed discussion of how this is an example of nonadaptive selection in the physical sciences.)

Fracchia and Lewontin recognize that a variational theory of evolution does not necessarily exclude transformational processes (e.g., aging) occurring in addition to the variational ones. However, our interest in this book is in cases where selective processes (variational processes) dominate in explaining the major features of change.

3.1.6 Selection and Causal Mechanisms

When we refer to an explanation in terms of a causal mechanism, we mean a well-articulated mechanism that takes an input and generates an inevitable, deterministic effect; a selection process can be such a mechanism (see appendix 10.4, the Price equation). Such a process repeatedly operates, producing predictable results over a specific timescale, and takes its output as input to the next iteration. Daniel Dennett (1995) called natural selection "algorithmic" for this reason; given the inputs, the influence of the mechanism on the system is more or less directly calculable.[11] Study of such algorithmic processes requires an assessment of the assumptions and logic underpinning their operation. In the case of nonadaptive selection, one of those assumptions is that of independence from local environmental conditions. When we say that we require a "causal mechanism" as a basis for the selective regime in a nonadaptive selection explanation, we are really saying that we must be able to explain causally the existence and prevalence of the selection regime, without recourse to environmental factors. Conventionally, we understand this to mean that we can actually specify a set of conditions, entities, and relationships that with the force of logic or mathematics can be demonstrated to act inevitably (on a particular timescale) to establish such a selection regime. Furthermore, we must be able to specify the entities that are subject to the selective regime; that is, we must be able to demarcate the domain over which the mechanism is valid and over which the selective regime is active. Included in the mechanistic structure are givens or assumptions about the various entities or relationships, which are subject to experimental or observational verification/falsification. Thus the mechanisms behave like any scientific explanations (or small-scale theories), varying in scope and certitude and being subject to refinement and falsification.

A causal, mechanistic explanation contrasts with *historical* explanations often employed in natural history. Such explanations, although recognizing that a variety of general mechanisms may be involved, are not made in terms of any specific mechanism(s), but rely for their explanatory power on *sequences* of realized historical events. Each event is historically unique and is itself derived from earlier events in the sequence. Such a derivation itself is the *historical explanation*, and the unique constellation of all relevant influences, as well as the unique result, means that purely historical explanations cannot explain general patterns across space and time. Some kinds of events may have happened in history more often than others, but historical explanation by itself offers little or no insight into why (see Nitecki and Nitecki 1992, and chapters therein).[12]

In the study of evolution as a whole, we use both modes of explanation. When our interest is in unique events, historical explanation or combinations of mechanistic and historical explanations are most useful. Explaining widespread patterns that are invariant across space and time, however, requires a mechanism that is also not dependent on local contingencies.

Our view of the structure of nonadaptive selection schemes (or for that matter, any selection scheme) emphasizes the key importance of the causal mechanism in making the scheme explanatory. The pattern or result is what needs to be explained. If all one has is the pattern and/or a history of selection, one still has no causal explanation. Simply identifying an apparent selection process does not suffice to explain the pattern that it appears to generate, since the causal nexus giving rise to the selection process itself is still unexplained. Selection (as opposed to drift or other random processes) must be caused by something, and that something has to be a general mechanism if a nonhistorical explanation is to be given. It is the cause of the selection process—the causal mechanism at its core—that explains the existence of the different probabilities of success for different configurations. So, for example, documenting the differential success or failure of different individual phenotypes, community configurations, or different species does not explain the observed pattern. Only with a mechanism is there a nonhistorical causal explanation. Only if that mechanism explains a standing selection regime is there an explanation of something that develops, arises, or is maintained beyond the timescale of each iteration of the mechanism (as long as conditions remain the same). And only if the mechanism explains a nonadaptive selection regime is there the potential for explaining, via selection, a quasi-universal, recurrent pattern across large spans of space and time—that is, for the explanation to function as an evolutionary or ecological "law" (see section 3.2).

3.2 Nonadaptive Selection Processes: Do They Generate Laws of Nature?

Traditional philosophical views saw science as being composed of a formal, pyramidal, logical structure involving laws and axioms from which statements about the real world were deduced. Contemporary philosophy of science largely rejects this view, replacing it with a variety of perspectives in which structures called *models* play a large or dominant part. (See appendix 10.9 for further discussion of theories and models in contemporary philosophy.) Models are defined differently by different authors, but generally refer to a substantial range of conceptual schemes and representations that link more abstract concepts (themselves perhaps expressed as models) to concrete empirical phenomena and provide the basis for explanation of those phenomena. Philosophical models in this sense may or may not refer to the same conceptual models and theorizing engaged in by working scientists, and they may include nonlinguistic objects such as diagrams, material models, or museum collections (Griesemer 1990b; Magnani, Nersessian, and Thagard 1999; Giere 2008).

So, if it is to be mostly "models all the way down," what happened to the *laws of nature* that used to play a major role in people's understanding of science? In the traditional view laws were required for scientific explanation, but in current understanding laws seem to have no necessary role (Giere 1999; Teller 2004). But that doesn't mean that no laws exist; if they do, it would seem that they would still be very useful, even if they are only one of the options available for making scientific explanations.

But laws, as traditionally defined, are in trouble (Teller 2004). It used to be thought that there were truly universal laws, at least in physics, but even there, strict, fundamental laws have come under suspicion—by philosophers and scientists alike (Cartwright 1983; Giere 1999; Laughlin and Pines 2000; Laughlin 2005). Even in mathematics there are no pure logical structures, as Gödel showed in 1931 (Raatikainen 2022). Biology might well be without such strict laws altogether (Brandon 1997). Cartwright (1998, 1999) famously referred to science's view of the world as being "dappled" and said that the laws that describe it are "a patchwork, not a pyramid." Nevertheless, "laws" are still of interest, since laws (or something like them) seem still to be used frequently by scientists in explanation and theory building. Even in biology there seem to be many generalizations of great scope and explanatory power that might be candidates for laws (Colyvan and Ginzburg 2003; Dodds 2009; McShea and Brandon 2010; Brandon and McShea 2020). Is there something specific that characterizes laws, or do we just get lucky sometimes, and some generalizations are on more solid ground than others?

Once again, philosophical views vary, but among those who see a role for laws in contemporary views of science a common set of criteria emerges from the discussion. First, if laws have any useful meaning, they are not *fundamental* laws, in the sense of exceptionless, precisely true universals (Cartwright 1983, 1999; Giere 1999; Ginzburg and Colyvan 2004; Teller 2004; Laughlin 2005).[13] Instead, four criteria are most commonly associated with a more pragmatic concept of laws:

1. *Necessity.* There has to be some sense in which the law is a regularity based on a set of necessary, inevitable relations among the things to which it refers, rather than on accidental associations or merely events that occur frequently.

2. *Domain specificity.* The law is operative only within a specific domain (defined circumstances or set of conditions) to which it applies.
3. *Universality*, within the domain. The law operates the same way throughout its domain.
4. *Ceteris paribus — all other things being equal*: The law describes only what would happen in an ideal case in which all confounding influences are absent.

These properties correspond directly to the properties of the nonadaptive explanations we discuss in this book. *Necessity* is given by the selection process, which distinguishes among phenotypes in their likelihood of persistence. A nonadaptive selection process has a specific *domain* of applicability within which it can operate. Within its domain it is *universal*, because the fitness differences upon which it is based do not result from interactions with the environment, but come from variation in inherent properties. And it describes what would happen in the absence of confounding influences (*ceteris paribus*) — not the actual, *joint* outcome of nonadaptive selection and concurrent adaptive evolution or other historically contingent events.

The domain of nonadaptive selection on community modules discussed in chapter 2 (three-species interactions in food webs) may seem relatively restricted and of limited importance. It may seem an exaggeration to elevate this process to a "law." But, the domain-specific universality of even this example is far more significant than it may at first appear. We have no doubt that should we find life on other planets, and should there be community food webs involving predation interactions, and should no extra-domain processes overwhelm the nonadaptive signal, those food webs will be dominated by the four most stable triplets. What's more, this should hold true wherever in the universe food webs are subject to the law (i.e., food web structures belong to its domain), throughout all time, past and present. All of the biological details will likely be different from anything we know from our own biosphere, but the pattern will be the same.

We cannot actually make a solid claim that any or all nonadaptive selective explanations are scientific laws, because there is no generally accepted definition for what a law is. But what we can argue is that nonadaptive selective processes exhibit the properties that post–logical-positivist philosophers most commonly ascribe to scientific laws. Nonadaptive selection processes match the prevailing concept of scientific laws without the need for stretching those concepts or for special modifications to the definition of laws, and nonadaptive explanations match law concepts much more closely than do other prevailing explanatory schemes (such as local adaptive processes), which are used to explain unique historical outcomes in biology. Laws, in the sense described here, cannot explain unique historical occurrences; if something happens only once, then it makes no sense to ask whether it is the result of a law or not. In order for laws to apply to historically contingent events, we would need to stretch the meaning of laws significantly (e. g., Brandon 1997; Mitchell 2002).

We regard nonadaptive selection mechanisms and the patterns they generate as resembling laws not because they are merely widely observed, or produce good predictions, or are useful in explanation, but because they are based on a kind of mechanism that can be studied directly and that generates a process and pattern that has no option other than to be lawlike. We think of biological laws (or lawlike statements) not as referring to things that happen statistically frequently, or to things that are universal properties of matter, but rather as universal mechanisms whose biological domains can be specified and that operate independently of time and space.

Our view that nonadaptive selection explanations are the basis for good laws or lawlike generalizations does not constrain the numbers of biological laws that there may be, nor does it exclude other sources of evolutionary and ecological laws. Nonadaptive selection mechanisms are not the only potential source of lawlike generalities in biology (Ginzburg 1986; Loehle 1988; Turchin 2001; Colyvan and Ginzburg 2003; Mikkelson 2003; Ginzburg and Colyvan 2004; McShea and Brandon 2010; Brandon and McShea 2020). However, we argue that any legitimate claim to a biological law must share with nonadaptive selection processes the properties of domain-specific universality and lack of dependence on time and space that characterize other *ceteris paribus* laws. We also expect that there will always be many biological processes that may be useful in explanation but that will not rise to the status of "laws." The lawlike generalizations with which we are concerned in this volume are about ecology and evolution, and they are generated by evolution through nonadaptive selection. Whether these are generally accepted as laws or not, we can *use* them like laws in explanation and to elaborate more extensive bodies of theory. As *ceteris paribus* generalizations pertaining to well-characterized domains, they are particularly robust.[14]

So, following Mikkelson (2003) and others, we usually refer to nonadaptive selection as generating "lawlike" generalizations. Rejecting the existence of truly fundamental laws, in biology or elsewhere in science, would largely remove the obstacles to calling nonadaptive explanations "laws." We personally lean toward calling them genuine laws, for that reason. However, we hope that we have clearly described the nature of nonadaptive explanatory schemes and can leave it to the reader to find a comfortable way to characterize their resemblance to scientific laws.

3.3 The Properties of Macroscale Multispecies Systems

This is not a book about a specific type of biological system or about a specific level of organization; rather, it is about a type of biological process. Perhaps surprisingly, "At what *level* does nonadaptive selection operate?" is not a fundamental question for us. We assume that nonadaptive selection occurs at many, or perhaps all, levels—but its relative importance in determining evolutionary change or maintaining repeated patterns may differ among levels. We concentrate on multispecies biological systems because we think that among them the patterns produced by nonadaptive selection will be most clearly seen and provide the most far-reaching explanations, in contrast to levels (such as the organismic) where adaptive selection is so common and intertwined with any nonadaptive selective processes that their separate effects are hard to disentangle. But in discussing multispecies biological systems, it does not matter to us whether the selection is on the persistence of entire systems, some of which survive and some of which disappear, or is among stable and unstable components or configurations of those systems, some of which survive and others disappear and thus cause the aspect or structure of the system to change. The key fact is that it is a nonadaptive selective mechanism, extended in time, that does the work of change, or of maintaining an equilibrium. We feel that an overemphasis on *levels*—deriving in part from the long-standing discussions of levels of selection, group selection, and hierarchical "expansions" of Darwinian adaptive theory—may obscure our straightforward message.

3.3.1 The Basic Hierarchies of Ecological and Historical Entities

Nevertheless, we do have to review briefly the "levels" and entities with which we are mostly concerned. We briefly introduced these in section 1.5, and we are now ready to say more about what they are, how they form a hierarchy of levels, and how they relate to adaptive and nonadaptive processes. There have been many hierarchical schemes for particular

Ecological Entities   **Historical "Genealogical" Constructs**

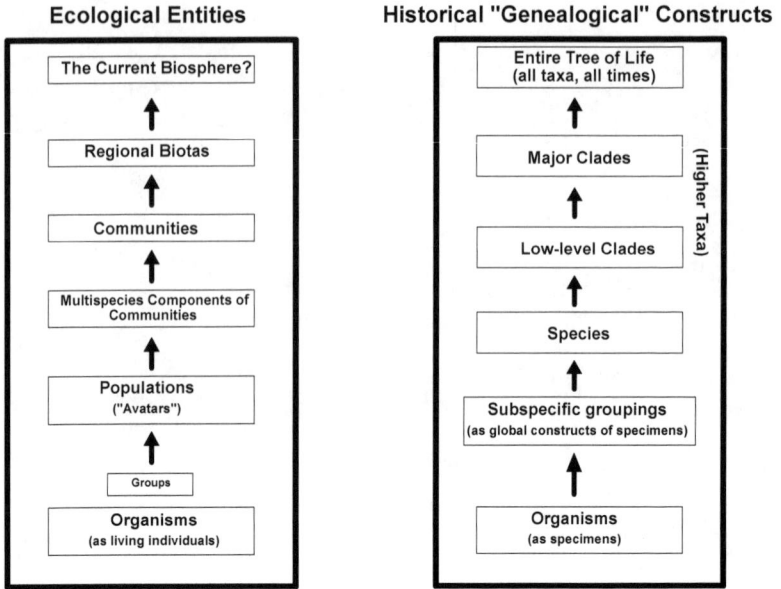

FIGURE 3.2. Commonly recognized ecological (*left*) and "genealogical" (*right*) hierarchies. Our version differs slightly from those published by Eldredge and colleagues, as explained in the text.

biological purposes, but the two major ones relevant to our discussion have been widely articulated (Eldredge and Salthe 1984; Eldredge 1985). They form a pair of quite different nested hierarchies, both beginning with the organism (figure 3.2).

Most hierarchical schemes focus heavily on the *objects* to be found in each hierarchy, even if those objects are not regarded as material. We think it more revealing of the nature of the hierarchies to concentrate on the *basis for membership* in each hierarchy.

On the left side of figure 3.2 are conventional, ecological entities that are recognized as spontaneously forming a hierarchy in nature. Organisms form local populations, those local species populations (or *avatars* of Damuth 1985) form multispecies components of ecological communities, they form communities, and so forth. We have indicated that there may be structure within local populations ("groups"); in the hierarchy there may also be "meta" and "sub" versions of some of the entities (not shown).

Membership in ecological entities is based, very broadly, on contemporaneous occupancy of a geographical location. Thus, ecological entities of a given level are able to interact with each other because they share a temporal and spatial context. These entities or ensembles exist "someplace," and we can visit them by going to a particular place on Earth. However difficult it may be to observe or measure the interactions among the components, nevertheless we are reasonably sure that each entity forms a material system whose components can plausibly interact with other components of the system. As such, the components not only interact within a specific local economy, but they also thereby share a specific environment. As we will see, this shared environment is crucial in the process of adaptation to local conditions by means of Darwinian natural selection. What happens in ecological systems is fundamental to the way that we ordinarily look at both ecology and evolution.[15]

Now consider the hierarchy on the right side of figure 3.2, often called the "genealogical" hierarchy. It represents a very different class of hierarchical entities—taxa—which we might call *global historical constructs*. As such they are a type of what we have identified as *global constructs* in chapter 2. In using this term we don't mean to imply that there is something unreal or arbitrary about taxa. Current usage regards taxa and their relationships as recording real historical events. Taxa are far more practical than ecological entities for writing the history of life, especially on the grand scale. But membership in *ecological* entities, such as organismal populations and communities, is based on concurrent occupation of a given place, forming a (potentially) interacting biological system. Membership in a species or clade is not at all the same; it is based on a network of historical genealogical relationships, like those of a family, and need not be restricted to a particular time and place. Like all such constructs, these entities are constructed by humans, even though they reflect events that humans did not control (e.g., speciation, extinction, clade expansion, etc.). Thus taxa, as historical constructs, do not represent ecological, biological systems, but rather sequences of historically related events whose components have no necessary overlap in time and space.[16]

"Organisms" appear in both hierarchies, but they are not the same thing when they do (Caponi 2016). In the ecological hierarchy, they are living (or recently living) individuals in given places, but in the hierarchy of genealogical constructs they are *specimens*. Specimens may be alive or dead at this moment but never cease to be part of the genealogies to which they belong (e.g., the specimens of *Tyrannosaurus rex* in the world's collections). Between specimens and whole species in the genealogical hierarchy there may be other identifiable genealogical entities (such as

subspecies) that record specific genealogical relationships, and we indicate these as a possibility.

Species are often treated as a special basal member of the genealogical hierarchy, because they and their presumed macroevolutionary
"populations"—which are usually taken to be (low-level) clades—are
sometimes hypothesized to undergo adaptive selection at their own level.
However, it doesn't seem auspicious to analogize taxa, including species,
with organisms or other ecological systems whose local populations can
potentially evolve adaptations to specific environments. Since taxa do not
necessarily occupy particular places or shared locations, and their members are usually not all contemporaneous, there is no obvious way that
taxa can adapt, by a process of selection *among taxa*, to local conditions.
There are no *local* conditions for taxa. There is no shared local population
for taxa. These considerations make it unlikely that, for example, *adaptive*
species selection is a meaningful concept, because there is nothing for the
species, as a whole taxon, to adapt *to*. Selection among such genealogical
units is discussed more fully in chapter 8 on macroevolution.

3.3.2 Multispecies Ecological Systems: Evidence for Selection
Is Not Evidence for Adaptation

Much of this book is concerned with selection among and within multispecies ecological systems—especially the ones in the middle of the ecological hierarchy such as communities or their multispecies components.
Some researchers dismiss discussion of communities as ecological entities because from the point of view of individual species (which are genealogical entities), the historical distribution of their populations among
individual communities is irregular and not entirely predictable. That is,
communities seem to be nothing but the chance overlap of individual species ranges (Whittaker 1975; Vellend 2016). Other researchers fear that
discussion of communities as entities reifies them as some kind of highly
integrated "superorganism" whose species populations (or the whole) can
evolve adaptively by "group selection." But between "chance association"
and "superorganism" there exists the possibility that we advocate, that
higher-level ecological units such as communities are primarily biological
systems of potentially interacting components that can, indeed, be subject
to *nonadaptive* selective processes.

Consider this conundrum: in spite of the fact that selection at these
ecological midlevels (communities and their multispecies components)

appears to be logically valid, there is little to no consensus on what constitutes an adaptation at these levels.

Selection seems to be well defined for these midlevel multispecies ecological entities. Selection is an extremely general process as described by
the Price equation (see section 3.1.4 and appendix 10.4). Like organisms,
these multispecies systems are "individuals" rather than "classes" (Hull
1980; Bouchard and Huneman 2013).[17] It is evident from the Price equation that reproduction (in the organismic sense) is not essential for *selection* to operate at any level. Likewise, the "heritability" that is required
by the Price equation is basically "like gives rise to like." If selection can
apply not only to organisms, but also to abiotic systems such as the solar
system (see appendix 10.5), to artificially constructed networks, to nonmaterial human constructs found in the genealogical hierarchy, to economics,
to crystal formation, and so forth (Price 1995 and appendix 10.8), then it
can certainly apply to midlevel ecological multispecies systems and their
components.[18] But contrary to the assumptions of many biologists, not all
ecological biological entities subject to selection can or are equally likely
to undergo *adaptive* selection.

The reason has almost nothing to do with the mechanics of selection
among these entities, but rather with the nature of the environment required for adaptive evolution. A key requirement for any kind of evolution
via selection is that there be heritable variation in fitness, *reliably* associated with a particular character or character state. In effect, each individual
in the population sharing an identical phenotype should be equally likely
to experience any fitness effects of bearing that phenotype. In ordinary
Darwinian adaptive natural selection, this reliable causal connection between expression of the phenotype and its effects on fitness is automatically forged because the individuals are in a *shared* environment; different
environments at other times and places are not relevant. Shared environments need not be homogeneous, as long as the members have an equal
chance to experience all aspects of the heterogeneity. (In more complex,
structured populations the fitness of an organism's phenotype is still known
within different contexts that the (meta)population provides; Wilson 1980;
Brandon 1990.) Occupancy of a shared environment is why, in adaptive
natural selection, the requirement is usually simplified to the question of
whether or to what degree the character alone is heritable, since ordinarily the fitness is "cemented" to the phenotype by the shared environment.

For *nonadaptive* selection, in contrast, fitness differences do not derive
from an interaction of the entities being selected and the local environment.

Significantly, there is also no difference, from place to place and across time, in the relationship between the character state and its implications for relative fitness, since those implications are based on intrinsic properties. Those properties do not change throughout the domain. So nonadaptive selection does *not* require a shared environment. Nonadaptive selection occurs within a specified domain, but not within a specific local environment. The association between phenotype and fitness effect is reliable throughout the *domain*.

Now we can see where the trouble begins. Researchers who have (convincingly) argued for the applicability of selection to higher-level *ecological* entities mostly assume that where there is selection (or differential fitness), there is adaptation (Bouchard 2008, 2011; Tran 2011; Doolittle 2014; Toman and Flegr 2017; Doolittle and Inkpen 2018; Lenton et al. 2021). Doolittle (2017) describes in detail his adaptive selective mechanism for higher-level ecological units, which is in fact a good description of the Price equation, except that he assumes that *all* selection is adaptive. Even those who consider these higher-level entities explicitly to be types of systems nevertheless lean toward the view that the value in studying them comes from detecting adaptation of the systems or adaptive evolution of their components (Toman and Flegr 2017; Lenton et al. 2021).[19] Often, *persistence* (or equivalently, stability) is treated as if it were always a kind of adaptation. Differential persistence can certainly be a component of fitness under evolution by selection. But persistence as an adaptation requires a particular causal nexus, as well as biological traits (novelties) that support both the relationship of persistence to a fit to some specific environment and the ability of variation in the characters underlying this fit to be further selected and build an adaptation. Harkening back to our remarks on selection for stability (see section 3.2), persistence alone is not evidence of an *adaptation* to local conditions.

Consider now the two cases of most interest to us, communities and their multispecies components (such as the community modules, or triplets, discussed in section 2.2). The overall community clearly forms a population and a local environment for the modules, but to what degree does it function as a shared environment for adaptive selection among the modules? For adaptation to occur, the modules would have to exhibit a fitness (stability, proliferation?) that is dependent on their interactions with the local environment, perhaps including other modules in the community. The modules, though, are not all necessarily freely experiencing the same environment; they are embedded in a network of interactions that has its own structure. This complicates the reliability of the relationship between component phenotypes and fitness, altering the way adaptations would be built and what local "adaptive challenge" is being addressed. At the higher, community level, it becomes even more difficult to specify what a shared environment for adaptive selection would be. If communities are defined as being coextensive with their habitats, then how do multiple communities share a habitat, and thus how are they able to exhibit adaptive evolution at their level?

Here and elsewhere, we do *not* claim that adaptive evolution at mid- to high levels in the ecological hierarchy cannot occur. Our book is not about adaptation, and someday a consensus might be formed that gives adaptation a clear and fruitful meaning at those levels. We do suspect that the lack of such a consensus now is indicative of the increasing difficulty of clearly formulating adaptive evolution at higher levels. It is tempting to conclude that the difficulty of defining adaptation for an entity is directly related to the difficulty of defining a shared environment for it.

Biologists have certainly been aware of the difficulty of explaining general patterns among higher ecological units by conventional, largely adaptive biological processes. *Coevolution* is an important and pervasive adaptive evolutionary phenomenon (Thompson 2005, 2013), but that very multifaceted pervasiveness makes it difficult to regard coevolution as a general process that just happens to drive specific, widespread, global regularities; there are too many possibilities for local adaptive coevolution instead (but see Solé and Bascompte 2006; Medeiros et al. 2018; Medeiros et al. 2021). Coevolution may lead to community components or communities that exhibit groups of interacting species that are highly mutually adapted. Yet if those coevolved species are within an interaction structure that is highly unstable with respect to nonadaptive selection, such a component will nonetheless be short-lived.[20]

In another direction, Mark Vellend (2010, 2016) developed a theory of ecological community dynamics and evolution based on the view that all patterns in community evolution can be described by four processes acting on species populations (avatars): selection (among but not within species populations, i.e., competitive outcomes), "drift," speciation, and dispersal. Vellend analogized these four to the elements of adaptive evolution by natural selection. There is of course always a reason that any one of these things happens. But even if all of the histories of the species populations are overlaid or synthesized in some way, this seems most unlikely to generate and explain an overall repeated pattern that emerges across communities at a higher level of analysis. Certainly it is possible to *describe* any sequence of events in terms of things happening to species populations, but this remains a description of a unique set of events—that is, a chronicle. Ideas, such as Vellend's—that the explanation of each case is tantamount to explanation of the overall pattern—transform the intended explanatory scheme into a historical one. That is, the theory permits only the historical explanation of a unique and contingent sequence of changes involving the particular species populations of each specific community. It is not a mechanistic theory of the community; it is instead merely a historical description of what happened to each species population. As such, it may be very useful in discerning important historical events, but there is no reason to expect history to add up to a limited and repeated set of overall patterns. A historical approach will not generate lawlike patterns that have the properties of nonadaptive selection processes, since it will not generalize across space and time. Higher hierarchical ecological entities such as regional or continental biotas share with communities many of the problems associated with adaptive evolution and although not well studied, tend also to be studied with a largely historical approach limited to species diversity and biogeography (Ricklefs and Schluter 1993; Rosenzweig 1995; Ricklefs 2004, 2015).

In sum, at and above the level of communities and their multispecies components, adaptive selection appears to have less explanatory power than at the organismic level. This is because it is difficult to define consistently the local environment of a community, the *local* population (of communities) that the community belongs to, and the way that "phenotypic" features of the community or its multispecies components contribute to its meeting adaptive challenges within the context of specific *local* conditions and potential competitors. Nonadaptive selection, on the other hand, is free of these handicaps. So the most important biological question to ask about a community is likely *not*, "How good is the *community's* fit to its local environment?"—an adaptive question—but rather, "What accounts for the stable presence of this community?"—a largely nonadaptive question. We argue in chapter 7 that communities can be seen generally to exhibit patterns resulting from nonadaptive selection mechanisms. This is because, as systems of ecologically interacting parts, their stability can be the result of nonadaptive selection processes that do not depend on the local environment. The effects of adaptive selection on ecological entities at that level are rare or absent. The upshot is the visibility at that level of the kind of domain-specific, universal, lawlike patterns explained by nonadaptive selection.

We hope to convince the reader that selection without adaptation is not a "failure" of evolution to accomplish something but may be interesting in its own right and, potentially, can accomplish much more of interest to ecology and evolution than might at first be suspected.

3.4 Summary

Given all of the uncertainty and controversy in the literature about when and where adaptation can be said to be occurring, it would be surprising if we could satisfy every reader with arguments about specific cases where it is *not* occurring. However, we believe that the features of quasi-universality shared by the kinds of explanations that we characterize in this book as based on nonadaptive selection processes endow these specific processes with uniquely powerful explanatory scope at the macroscale.

In our view, there is no single "theory of nonadaptive selection," with a grandeur and sweep that rivals that of a general theory such as neo-Darwinian natural selection. Rather, we see nonadaptive selection as a *conceptual framework* embracing a heterogeneous set of mechanisms that share a key set of features. First, these explanatory schemes are about patterns that are the result of selective processes, which means that the theory translates short-term causal processes into their implications for long-term selective outcomes. Second, the traits whose variation yields differences in fitness are fundamental, intrinsic properties of the systems or entities in the domain of each nonadaptive selection process, and thus the selection regime so defined is not the result of an interaction between the system and its local environment. Thus these mechanisms are nonadaptive; acting alone they cannot adapt their system to its local environment. On the other hand, this means that the selection regime of a given nonadaptive process applies in the same way to every instance in its domain. A nonadaptive selection process thus can explain widespread, recurrent patterns in space and time. Significantly, the systems under comparison do not need to form ecological populations. What is more, nonadaptive mechanisms can thus be considered to generate lawlike statements about their domain. To the degree that other processes do not overwrite the effects of nonadaptive selection, those effects will be ubiquitous. The patterns explained by nonadaptive selection theories thus have implications for the question of whether there are laws in ecology and evolution, and what those laws may be like.

CHAPTER 4

Nonadaptive Selection at the Level of Single Species

4.1 Fisher's Principle: The Maintenance of Even Sex Ratios

Ronald Fisher (1930) advanced an explanation for the widely observed 1:1 sex ratios among sexually reproducing species;[1] *Fisher's principle* is one of the most well-known generalizations in evolutionary biology. Briefly, Fisher argued that when there are more individuals of a given sex in the population, selection favors individuals that produce more of the opposite sex. This produces continuous selection toward an equilibrium, 1:1, sex ratio.

The selective process of Fisher's principle operates the same way in any and all environments external to the population; it does not operate differently if the local environment changes and does not depend on the specific adaptations or characteristics of the species, or even upon how sex is determined (other than the presence of two sexes). From this we recognize that the Fisher process itself is an example of *nonadaptive* selection at the level of populations within a species.

The fitness differences upon which the Fisher mechanism is based are those among individuals having a tendency to produce different, uneven sex ratios among their offspring. Many evolutionary biologists claim that the Fisher mechanism is just a form of *frequency-dependent selection* or an aspect of *parental investment*, usually to make the Fisher process integrate better with a large number of simultaneously operating processes of ordinary adaptive selection that represent various adaptive evolutionary scenarios for sex-ratio evolution (Bull and Charnov 1988; Hamilton 1967; Carvalho et al. 1998; Sober and Wilson 1998). But in making this claim for the Fisher process itself, they make an unusual conceptual maneuver. The "real" environment of the population, whatever it may be like, is, *for the purpose of the Fisher process*, discounted and replaced by nothing more than the frequencies of the opposing sexes. So now the action *of the Fisher process* depends only on the state of a system; the relevant fitness differences among individuals depend only on the current configuration of the population (its sex ratio), not on an interaction between individuals of either sex and aspects of the local external environment. Thus, ironically, this reformulation isolates the Fisher mechanism from adaptive evolution and makes it one in which the relevant fitnesses do not "see" the local environment. That is, the Fisher mechanism is now an example of a nonadaptive selection process, which is exactly what we claimed at the outset. Populations never "adapt to" or "solve the problem of" the sex ratio via the Fisher process; rather, they always experience this nonadaptive selective process, always in the same way, and thus frequently exhibit the equilibrium value, a 1:1 primary sex ratio—as long as *other* processes (which might or might not be adaptive) do not overwrite its action. The entities experiencing low fitness due to Fisher's process alone are not failures at dealing with the local environment but are the entities selected against in an unstable configuration of the system.

Sober (1984, 2000) argues that Fisher's principle is a genuine "law" of biology. His argument parallels both our depiction of nonadaptive selection and our discussion of biological "laws" in section 3.2.

As it happens, this case of sex-ratio evolution by nonadaptive selection also illustrates another of our points: how easy it can be at the organismic population level for adaptive evolution to overwhelm or to co-opt a nonadaptive selective process. Sex ratios in nature are not all 1:1; the existence of a degree of population structure makes it possible for local adaptive selection on the primary sex ratio to override the nonadaptive selection underlying the Fisher mechanism (Hamilton 1967; Sober and Wilson 1998, 38–43). Thus, the sex ratio expected from nonadaptive selection may frequently be obscured by adaptive evolution. Examples like this suggest that it often will not be useful for biologists working at the organismic level to distinguish nonadaptive from adaptive selection, because of the ease with which adaptive selection can operate and the likelihood that any specific case reflects the action of *both* adaptive and nonadaptive processes. Conversely, among and within systems where adaptive selection processes are absent or weak (such as macroscale multispecies systems), the patterns produced by nonadaptive selection should be more apparent and widespread.

4.2 Nonadaptive Selection Explanation of Heterozygote Fitness Superiority

In populations of outcrossing diploid animals and plants we commonly observe that heterozygotes have higher fitness than the corresponding homozygotes. This *heterosis* (or *hybrid vigor*) is referred to in most textbooks on evolution. The flip side of this observation is the finding that inbreeding, which increases homozygosity, often causes a decrease in fitness. So if in most of the instances when we observe heterozygotes we see that they have higher fitness than homozygotes, this raises the question of whether there is something generally superior about heterozygotes—something that allows them to confer fitness advantages whatever role the alleles at that particular locus play in adaptive evolution.

In the past there were attempts to suggest mechanistic explanations for general heterozygote superiority (Lerner 1954; Fincham 1972; Clarke 1979). However, we believe that the reason heterozygotes usually have higher fitness than homozygotes is that selection removes the alternatives. Unfeasible or unstable genetic equilibria create nonadaptive selection, acting on the fitness values to preserve cases in which heterozygotes are superior and to remove the others.

We remind the reader that in the textbook case of one locus and two alleles, this would have been an easy conclusion. If the heterozygote (AB) is superior to both (AA) and (BB) homozygotes, the locus will reach a stable genetic polymorphism, with both alleles A and B present in certain frequencies. The reason for this stability is that heterozygotes, when crossed, still produce 1/4 AA and 1/4 BB homozygotes—the heterozygotes cannot win and achieve a frequency of 100%. With any other arrangement of fitnesses (with heterozygote inferior or intermediate in fitness relative to the homozygotes), one or the other of the two alleles will take over the population. At the resulting monomorphic locus we will see no heterozygotes. So, if we see any heterozygotes at all, it has to be the case that they are superior in fitness. The well-known example of sickle-cell anemia in humans is such a stable polymorphism. Homozygotes for the sickle-cell mutation are seriously ill, but the heterozygote is almost asymptomatic and in addition shows resistance to malaria. Thus in malaria-infested areas the homozygotes both have lower fitness than the heterozygote, and selection cannot remove the sickle-cell allele from the population.

However, analysis of cases of multiple alleles at one locus, or multiple loci (in which gene interactions across loci complicate the situation) is not so straightforward. Nevertheless, we can show that the same logic applies in general, where nonadaptive selection accounts for the prevalence of heterosis by removing alternative configurations.

The more complex cases can be analyzed by computer simulation, beginning with the case of multiple alleles at a single locus. Let us first consider two alleles, since we already know the answer. Suppose we have the computer generate a large number of 2 × 2 fitness matrices, assigning the three genotypes (AA, AB, BB) fitnesses at random. We ask the question, "What is the proportion of cases that exhibit a stable polymorphic equilibrium?" (We answer this by evaluating the properties of the matrices. There are three possibilities: stable equilibrium, instability, and a stable polymorphism that is nonetheless unfeasible—i.e., requiring negative frequencies.) We know from the preceding discussion that a stable feasible equilibrium occurs only when the heterozygote is superior in fitness to the two homozygotes, and by simple probability, when we generate the random 2 × 2 matrices, that should happen one-third of the time. Not surprisingly, in the computer runs, one-third of the matrices are feasible and stable, and all of the heterozygotes in those cases have superior fitness to the corresponding homozygotes.

We then iterate the same procedure, each time adding another allele. The results show that for three or more alleles attaining a stable equilibrium is very difficult; a simulation of six alleles showed no stable equilibria at all in a sample of 100,000 matrices (Lewontin, Ginzburg, and Tuljapurkar 1978; see figure 4.1).

The Lewontin, Ginzburg, and Tuljapurkar (1978) article is known for showing that selection can maintain only a few alleles in a polymorphic locus. The seldom-noticed connection is that this is a direct result of nonadaptive selection among fitness configurations. This was certainly not the language used at the time of publication of that article, but the observation, in fact, was possibly one of the first examples of the power of nonadaptive selection.

Even if one forces every heterozygote to be superior to its homozygotes, the proportion of stable matrices still drops rapidly as the number

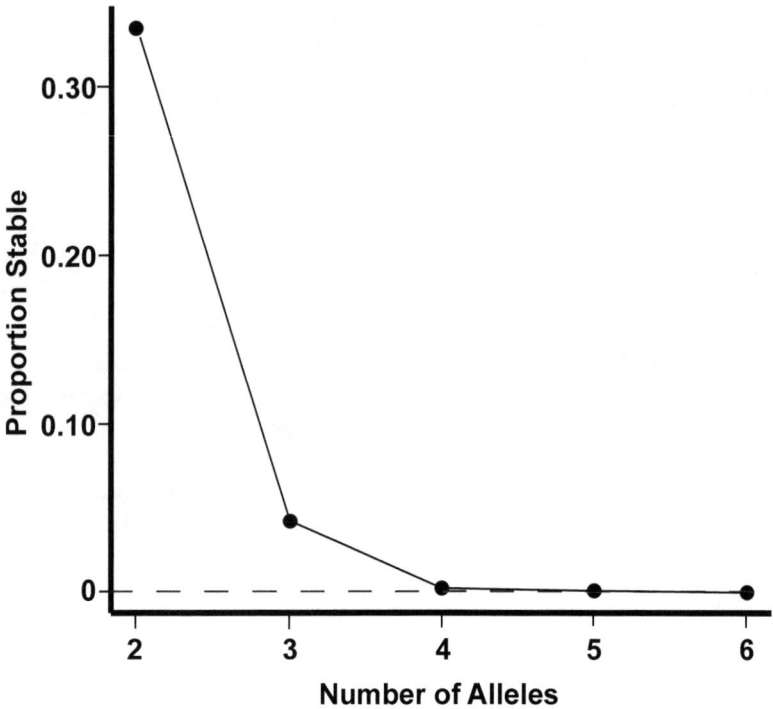

FIGURE 4.1. Chances of stable polymorphism with various numbers of alleles and initially fully random fitness assignment. Figure after Lewontin, Ginzburg, and Tuljapurkar (1978).

of alleles increases. So we will not see very many multiple-allele stable polymorphisms. But when we do see them, are all of the heterozygotes superior to the homozygotes? It turns out that the criterion for a feasible stable equilibrium with more than two alleles is merely that the average heterozygote in such a system be superior to the average homozygote (Ginzburg 1979; see appendix 10.10). Even so, this still leads to the conclusion that we have to expect to see heterosis widespread, on average, as the result of feasibility/stability selection.

An analogous result is found for multiple loci. Michael Turelli and Lev Ginzburg (1983) assigned fitnesses at random to multilocus genotypes and ran selection simulations until each one stopped evolving (either reaching a feasible stable equilibrium or becoming monomorphic at all loci). They then compared the fitnesses of monomorphic genotypes with genotypes having heterozygotes. In the majority of the cases, monomorphic genotypes were less fit than genotypes having one locus that was heterozygous, and genotypes with two heterozygous loci were even more fit than those having one. Thus, the single-locus results are supported by these multi-locus generalizations, another example of nonadaptive selection.

In summary, what looks like a general property of heterozygotes to exhibit high fitness is instead simply the result of our having mostly the high-fitness heterozygotes available for observation. The others have been removed by selection because they do not form feasible and stable equilibria. New mutations always appear in a heterozygote condition but in most cases are of low or intermediate fitness and are quickly removed; monomorphism is a more common case than polymorphism. No special explanation for the prevalence of heterozygote superiority is necessary. Widely observed heterozygote superiority is a biological, lawlike generality stemming from nonadaptive selection.

The sickle-cell heterozygote is certainly an adaptation to an environment in which malaria is present. There probably are similar adaptive stories for every single-locus, two-allele, stable polymorphism. With multiple alleles and particularly with multiple loci, it will not be easy to come up with a simple verbal explanation. A general heterozygote superiority law or, in other words, commonly observed hybrid vigor, is clearly a result of the elimination of other alternatives through nonadaptive selection. But heterozygote advantage as a general "law" is not an adaptation to anything in particular.

CHAPTER 5

Nonadaptive Selection Explains the Size Scaling of Lifespans

5.1 Introduction: The Problem of Lifespan Scaling

We now pick up the thread of our discussion of ecological allometries, which we introduced in section 2.3. There we argued that the interspecific energy equivalence relationship arises from a nonadaptive selection process, and this is why it is one of the lawlike generalizations that we observe widely in nature.

But there are a large number of other ecological allometric relationships known throughout ecology (Peters 1983); are they all examples of different nonadaptive processes and examples of their own independent ecological laws? Surely not all are. As chapter 6 shows, most allometric relationships are easily explained as algebraic consequences of (or simple deductions from) a smaller number of other, more fundamental relationships (see also Lindstedt and Calder 1981). The interspecific scaling of lifespan is of particular interest, however, because although lifespan tends to scale with body size with the same exponent as do other, physiologically based biological times, it is unclear why it should do so. In this chapter we focus on how the scaling of lifespan can be explained as a result of a nonadaptive selection process.

For concreteness, in our examples we use the "quarter power" exponents that tend to be widely observed but are by no means universally accepted, nor are they likely to be precisely seen in a given empirical dataset. Our argument in this chapter and in chapter 6 does not depend on the quarter-power exponents, only on the fact that lifespan variables have the same scaling exponents as do the physiological times.

5.2 Traditional Explanations

5.2.1 The Identity of Scaling Exponents of Lifespan and Physiological Times: Is It Just a Coincidence?

One large set of biological allometries involves physiological and life history traits that are expressed as rates or as times. Rates refer to the frequency of cyclical or repeated events per unit time. Generally, physiological rates (e.g., mass-specific metabolic rate, heart rate) scale not far from body mass to the $-1/4$ power. Times, on the other hand, are just that, the time that it takes for events to happen (e.g., time between breaths, gestation time, age at maturity). Physiological times usually scale reciprocally to the rates, as mass to the $+1/4$ power. Often these scaling relationships seem straightforwardly related to those for metabolism and to the energy that must be expended to perform the subject activity or cycle (Lindstedt and Calder 1981; Calder 1984; Charnov 1993).

But there are other life history variables, such as lifespan and generation time, that are not primarily physiological variables, in the sense that they are not involved in the day-to-day functioning of the organism. Rather, they describe how members of the species function together in an ecological context, at the population level, and over long spans of time. Yet these largely nonphysiological variables *also* scale interspecifically with body size the same way physiological times or rates do, that is, not far from the $\pm 1/4$ power of mass (Lindstedt and Calder 1981; Peters 1983; Calder 1984).

This remarkable fact seems to demand a causal explanation. It is not obvious why physiological functioning per se should also determine, say, the number of births across the lifespan, which has population and demographic implications that may not be wholly captured by the relative physiological performance of an individual. The physiological times may explain the individual reproduction rate per hour, day, or year, since reproduction is a physiological process. But it is these other, more "ecological" variables associated with longevity, such as lifespan and generation time, that must scale in such a way as to permit viable ecological populations.

So the conundrum is this: lifespan scales in such a way that it is *consistent* with metabolic rates and the day-to-day functioning of the organism.

But if the organism functions well, why should it not live forever, and if not, why should it live a span of time apparently dictated by metabolism? Does the coordination of longevity variables with physiological ones represent some particular causal process, or is it just an astonishing coincidence?

Traditional theory of the evolution of life history posits that at the level of natural selection among individuals, optimizing relative fitness involves balancing numerous trade-offs among different life history traits. Thus the evolution of any one trait is thought to be rigidly constrained by the other trait values, which includes the way that they scale with size. However, traditional life history theory cannot apply directly to the scaling of longevity variables. Unless there is a causal link between the scaling of physiological variables and that of longevity variables, there is no sure basis for a trade-off between them.

5.2.2 Explanations Based on Aging Processes and the Accumulation of Damage

Consider a compound variable, such as the amount of energy a given volume of tissue metabolizes over a lifetime. This variable is the product of lifespan, which scales as +1/4 power of body size, and mass-specific metabolic rate, which scales as −1/4 power of body size; lifetime tissue metabolism should thus scale with body size as $1/4 - 1/4 = 0$. It will be *invariant* with respect to body size (see section 10.1.1). Individuals of different species seem to metabolize the same total amount of energy, per gram of tissue, during their lifetimes, regardless of size.[1]

This invariance of lifetime energy use for a given volume of tissue has long intrigued researchers. Max Rubner (1908) first recognized it and realized its implication: that the tissues of all species appear to "wear out" at the same rate; this, in turn, should determine the scaling of aging phenomena and, from that, maximum lifespan. For Rubner, it appeared that simply by metabolizing, tissues and cells either use up a limited store of some capacity to continue functioning or accumulate damage from processes proportional to the total metabolic flux they have experienced. The former possibility was the basis of an early "rate of living" theory (Pearl 1928), but in its literal form that theory is contradicted by numerous empirical inconsistencies and is no longer thought to be an adequate basis for explaining either aging or lifespan (Austad and Fischer 1991; Speakman et al. 2002; Speakman 2005; Austad 2010). On the other hand, the accu-

mulation of damage has stimulated wide interest in the form of a number of theories and has the advantage of drawing a more or less direct
connection between metabolic rate and aging phenomena. Such theories
posit that aging is the result of tissue damage accumulated over time from
metabolic activity. This might be due to the production of free radicals
and oxidants (Harman 1956; Sohal and Allen 1990) or to the deviation
from molecular steady states (Sacher 1959; Demetrius 2004; Speakman
2005). Despite widespread interest in accumulation of metabolically related damage as a theory of aging, it too suffers from an increasing array of empirical contradictions or uncertainties as a general explanation
of life history phenomena such as lifespan (Wickens 2001; Selman et al.
2012; Glazier 2015; Brown, Hall, and Sibly 2018). Furthermore, even if
accumulation of damage has merit, it still does not really answer the question about coincidence that we are asking. Why should the various *aging*
processes, in the face of all the potential impacts they have on the fitness
of individual organisms and thus all of the opportunities for selection to
alter them, remain so closely linked to metabolism or other metabolically
linked functions?

5.2.3 The "Long-Term Average" Interpretation

An alternative to cellular- and tissue-based explanations for lifespan scaling is what might be called a long-term average (LTA) explanation. Simply stated, an LTA argues that most populations of species are, most of
the time, near a population size (K) dictated by the amount of resources
available to them. At that equilibrium point, K, the population is neither
growing nor shrinking, which means that the average individual just re-
places itself. Thus, at equilibrium the scaling of lifespan and generation
time will reflect primarily the length of time it takes for the average adult
to produce one offspring that lives to adulthood (Ginzburg and Damuth
2008). The time it takes for an individual to produce an offspring can be
estimated from the rate of energy expenditure, growth rates, and reproductive schedules of individuals, and thus the scaling exponent of the time
to produce this offspring will tend to conform (proportionally) to the scaling exponent of these physiological and life history traits.

James Brown and colleagues (Brown, Hall, and Sibly 2018; Burger,
Hou, and Brown 2019; Burger et al. 2021) elevated an LTA explanation to
a general principle (the "Equal Fitness paradigm," EFP).[2] They confirmed
that across species of all body sizes, a unit of biomass really does tend to

expend the same amount of summed energy, over its lifetime, as Rubner thought. This sum is not very different for different species, in spite of large variation in life history details. Thus the proportion of total energy expenditure of species or individuals that is represented in the offspring of the next generation is seen empirically by the EFP to be highly constrained across body sizes, but that constraint itself is not explained (see section 5.3; Brown et al. 2018).

Populations will always equilibrate at K, somehow, but that does not in itself constrain the degree to which the scaling of any one trait may evolve to wander away from the scaling for the average values of physiological rates and times. Although in an LTA explanation there is a complete consistency of the scaling of traits at all levels, from physiological rates to lifespan, the implication is that the upper and lower bounds of the allowed evolutionary excursions *happen* to be symmetrical and will tend to "center" the average exponent values of all traits across all species, such that they remain near those exhibited by physiological variables. Again, what we see is that everything in the LTA explanation works, because everything is just so; should we trust an explanation based solely on such a coincidence? Or is there some process that actively keeps the whole structure centered on scaling exponents that are characteristic of physiological times?

5.3 Physiology and Ecology Interact to Provide a Causal Explanation via Nonadaptive Selection

5.3.1 Population Dynamics Is Critical

Our view is that the causal process that links the scaling of lifespan variables to physiological ones is a specific nonadaptive selection process acting on populations, with population growth rate as its target. LTA theories do not articulate a clear nonadaptive selection mechanism, because in their view all populations will reach K and more or less stay there. All species end up in the same place, but there is no knowledge of how they may have gotten there. Another way to put it is that LTAs are not causal explanations themselves (and almost certainly aren't Kuhnean "paradigms"), but may be useful sets of observations that need explanations.

Even if all species are at equilibrium (K), selection on individual *relative fitness*, in the form of selection for greater individual production of offspring, can occur within populations every generation (given appropriate variation, etc.). Such individual selection does not necessarily result in a change in K, but the reproductive behavior of the individuals composing the population at equilibrium will have evolved as a result. So although K need not change as a result of this evolution, the *dynamics* of the population's recovery from low population sizes and its behavior as it reaches or exceeds K can change dramatically (see section 5.3.2). Thus, the mere existence of K does not maintain correspondence of the scaling of the durations of longevity variables and physiological times, because it does not counteract this tendency for individual reproductive capacity to evolve or, for that matter, any evolutionary change happening within populations that affects population dynamics.

5.3.2 Density-Dependent Population Growth and Chaotic Dynamics

Discrete difference equations are often used to model the dynamics of single-species populations. In general these models incorporate reproduction, mortality, and some form of density dependence (broadly, the depressive effect on population growth rate as the number of conspecifics using up resources increases). In such models, populations grow at their maximum rates (R_{max}) when the population is low and there are no density-dependent effects, and stop growing if they reach a stable equilibrium carrying capacity (K; see section 7.2.1).

We denote by R_{max} the net reproductive rate for a population in the absence of density-dependent feedbacks; it represents the maximum potential growth rate of that population *per generation*. An equivalent and intuitive life history interpretation of R_{max} is that it is the number of successful offspring of the average female, per generation.

Robert May (1974) showed that, surprisingly, population dynamics of even the simplest such models become chaotic when a population's R_{max} exceeds a certain threshold value. Populations whose reproductive rates exceed this *May threshold* (Ginzburg, Burger, and Damuth 2010) are likely to be prone to extinction, because their numbers fluctuate wildly and unpredictably (May and Oster 1976; Thomas, Pomerantz, and Gilpin 1980; Berryman and Millstein 1989). High levels of chaos are seldom seen in nature (Pool 1989), though low levels may not be uncommon (Rogers, Johnson, and Munch 2022). Time series data suggest many populations exist close to, but not over, the threshold of chaos (Doebeli and Koella 1995; Ellner and Turchin 1995). This is the basis for the nonadaptive selection process we identify here: *populations that evolve too high a population growth rate* are unstable and will be removed rapidly by extinction such that they simply will not be observed (see Ginzburg, Burger, and Damuth 2010 for additional details; in that publication we referred to nonadaptive selection as "ecological elimination").[3]

This nonadaptive selection mechanism implies a general allometric regularity. A given model of density-dependent population growth defines quantitatively how population growth rates change with population size. This can be thought of as a curve of a particular shape. Different species or higher taxa will likely have somewhat different curves, but it is reasonable to assume that adaptively and ecologically similar species will have similar curves. Remember that R_{max} is measured as a number per unit of generation time. This means that however generation time might scale with body size, organisms of different sizes (generation times) that share the same density-dependent population growth model (i.e., have the same-shaped density-dependence curve) *have the same May threshold*, that is, they will transition to chaos *at the same R_{max} value* (Ginzburg, Burger, and Damuth 2010). If R_{max} is limited to values near to but not exceeding the May threshold, then for any group of similar species sharing roughly similar density-dependence curves, there should be *no* relationship among species between body size and R_{max}. Recall that R_{max} is *also* the number of successful offspring produced by the average female per generation. Therefore, *this number of offspring* should also be constant across body sizes (again, given similarly shaped density-dependence curves).

Empirical data suggest that this is the case. The body-size invariance of R_{max} has been recognized to apply to vertebrates since at least the early 1980s (Calder 1984). Eric Charnov, Robin Warne, and Melanie Moses (2007) showed that it is true for lizards. Plots from Lev Ginzburg, Oskar Burger, and John Damuth (2010) of estimated R_{max} versus body mass for mammals, birds, and fish also showed body-size invariance (see figure 5.1). Within these taxa, across the range of their body masses, there is little or no relationship with R_{max}. The mean R_{max} values for mammals (3.2) and birds (2.7) are similar, but that for fish is higher (17.0). This suggests that fish differ from birds and mammals in their density-dependence curves but among themselves still exhibit body-size invariance of their R_{max}. The variation of R_{max} seen *within* groups presumably reflects variation in the population dynamics of different species, as well as the difficulty of estimating R_{max} accurately in natural populations. The absence of any significant interspecific trend across wide ranges of body size is striking for the allometry of a life history variable.

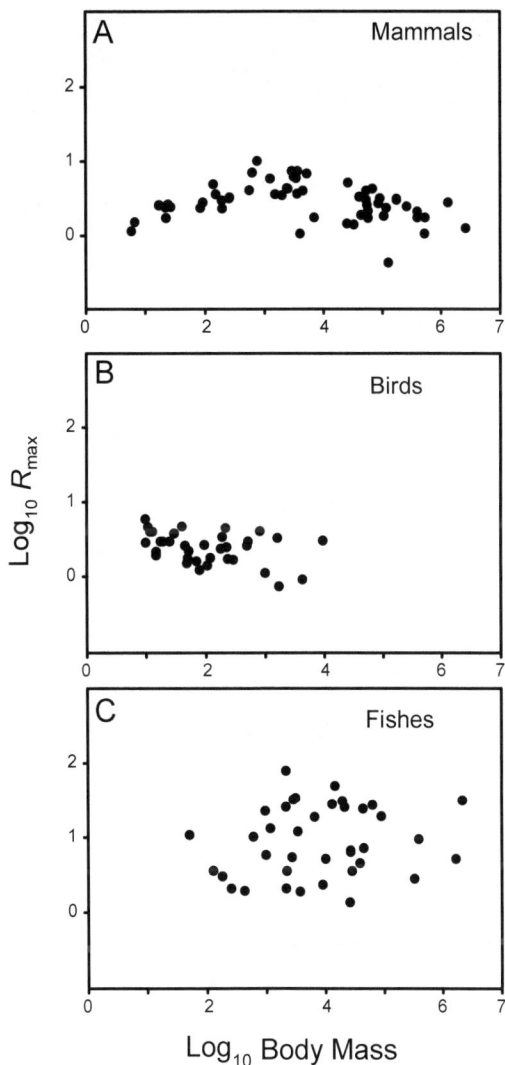

FIGURE 5.1. Maximum generational growth rate (R_{max}) versus body mass (B, in g), for groups of vertebrates spanning a large range of sizes and for which species R_{max} values are available from the literature. Both variables are log10-transformed, since our prediction is that the ordinary least-squares slope of such a relationship will not differ from 0. Across these three independent tests, in spite of differences in elevation, there is no significant departure from a slope of 0. Figure from Ginzburg, Burger, and Damuth (2010).

May's (1974) results were based on models with nonoverlapping generations. The same kind of instabilities have been shown to appear in more complex, age-structured models with overlapping generations, just as they do in the simpler ones (Levin and Goodyear 1980; Tuljapurkar, Boe, and Wachter 1994).

This invariance now allows us to explain the connection between the scaling exponents of physiological and longevity variables. With respect to their genetic contribution to future generations, individuals within populations always benefit from increasing their reproductive rate. This will lead populations to evolve to increase R_{max} as much as possible. But if the upper limit of R_{max} is near to but not exceeding the May threshold, then R_{max}—the number of offspring successfully produced by the average female—will be the same for any generation length. Put another way, the May threshold, expressed in terms of R_{max}, represents how many individuals can be produced in the average lifetime (generation time) of a species of any size while still keeping population growth from crossing into unstable territory. But offspring individuals can be produced only so fast. The rate at which an individual can produce offspring scales similarly to physiological *growth* rates (both scaling with body size interspecifically as −1/4 power). This is because reproduction involves growing new tissue, which is by definition determined by a species' average physiological growth rate. How much new tissue? For populations that are viable, we recognize the *ecological* requirement that lifetime offspring production (R_{max}) in any group of similar species be invariant with body size. This in turn means that longevity variables (such as generation time and lifespan) will tend to scale with an exponent near +1/4, as does the time it takes to grow a given amount of tissue. In other words, to maintain invariance in R_{max}, as body mass increases among species, the reduction in the physiological growth rates must be compensated for by an increase in lifespan and generation time.

Thus, the connection between the scaling of the physiological variables and longevity variables is here: the generation time and lifespan will both scale with body mass, as does the length of time it takes to produce R_{max} number of offspring, with an exponent of 1/4. The mechanism that opposes continued evolutionary divergence away from this central relationship in the scaling of all of these variables is a nonadaptive selective one: ecology interacting with physiology, through the dynamics of population growth rates, causing the removal of populations showing extremely low or extremely high population growth rates. This lock is not the result of a static equilibrium; rather, it is the dynamic equilibrium of a nonadaptive selection process that holds the connection in place.

This easily explains why, in spite of large observed variation in mortality and birth rates among taxa, the body-size scaling of longevity variables is so regular. There is no single, physiological life history character controlling the scaling of longevity variables with body size. Instead, any combination of life history traits that results in a viable population growth rate, high but not too high, leads to approximately the same scaling relationship.[4]

Furthermore, if one converts the R_{max} values into energy expended in the production of those offspring individuals, one sees the fundamental observation of Brown et al. (2018) that this energy quantity is conserved across body sizes. Our explanation thus does not contradict LTA observations or the EFP (or even the damage accumulation explanations), but rather explains why the patterns of scaling exponents they identify aren't fortuitous; similar amounts of lifetime energy expenditure will tend to characterize a wide variety of species of different sizes, because the ecological constraints on population growth rates are largely independent of size (within major groups). We regard the well-defined nonadaptive selection mechanism created by the May threshold, and the resulting body-size invariance of R_{max}, as a causal explanation for the widespread 1/4 scaling of lifespan. Notably, because this scaling generates a general, lawlike allometry by linking ecology and physiology, it appears to have an important role in structuring the overall scheme of interconnected allometric relationships in physiology, life history, and ecology. We now turn to this topic in chapter 6.

CHAPTER 6

Ecological Allometries, Nonadaptive Selection, and the Coordination of Powers

6.1 What Do Ecological Allometries Mean?

Clearly many traits that scale with size are interrelated, and there cannot be a separate causal explanation for the scaling exponents pertaining to each trait. How, then, do we distinguish fundamental allometric relationships from those that are secondary consequences of those relationships, and how are they related to each other?

A body-size allometry represents one or more quantities that change in a regular way across a range of body sizes. In functional, causal terms, this allometry must actually be the outcome of some constraint or process that works identically across size, such that it constrains the small and large-sized species or individuals to conform to the same relationship. Therefore, identifying the feature that is conserved or kept uniform across body sizes (i.e., something that would scale independently of size, as mass raised to the 0 power, and which we can also describe as being *invariant* with respect to size) yields a strong clue to the fundamental cause and biological significance of a given allometry. As we've seen in section 5.2.2, observation of invariance alone does not necessarily indicate the cause of an allometry, but it is likely that the fundamental cause is an invariance of *something*. That something need not be a simple trait and could also be the expected outcome of a selective mechanism (as in the interspecific scaling of energy use in chapter 2).

Before presenting our main argument in section 6.2, we want to more fully examine aspects of invariant properties in allometric scaling and the role of nonadaptive selection in allometric "laws." We begin with two well-known models that purport to explain universal scaling exponents of metabolic rates in organisms. Of course, as is usual at the organismal level (chapter 4), the actual realized metabolic rate of any species is the result of some degree of adaptive modification, which fits the organisms to specific ecological roles and environments. However, the claim of these two models is that there is nevertheless a general ("lawlike") tendency to exhibit a particular scaling exponent of individual metabolic rate with body size, based on some model that applies to all or most species and that applies across large samples of species in spite of their local adaptations (Savage et al. 2004). We recognize that at present this topic is controversial, and our intent here is not to argue for any particular model, or for how many models might eventually be needed to satisfactorily explain metabolic scaling. We are making a different point: that if there are such universal (or quasi-universal) regularities in the scaling of metabolism, at any scale, the explanation is built on an implicit argument for nonadaptive selection on some invariant functional constraint or property. We use the metabolic models to support this interpretation.

Historically, the earlier model is the surface rule, or Max Rubner's (1883) rule, which states that in order to dissipate heat produced by metabolic activity, it must be lost by diffusion across the body surface. Therefore, the heat production (metabolic rate) can increase no faster than the surface of the body increases, as the 2/3 power of body mass. The functional requirement that is preserved at all body masses is the capability of diffusing internally produced heat. The metabolic rate is matched to the surface area and should scale with body mass with an exponent of 2/3. If this were not so, even at rest the organism would continue to heat up or cool down indefinitely.

The second model concerns the internal metabolic transportation network for organisms that supply metabolites to their tissues from a source, such as those with a heart and circulatory system (West, Brown, and Enquist 1997, 1999; Banavar et al. 2002; Banavar et al. 2010). Here, it is the functional requirement that the metabolite transport system be able to provide the same rate of metabolite delivery to every part of the body, regardless of distance from the source, and without altering the blood/body-mass ratio, that ultimately gives rise to the "canonical" interspecific scaling exponent of 3/4 in the scaling of metabolic rate with body mass (in the most efficient networks). The other metabolically related variables all change with body mass in such a way as to preserve this compound constraint across body masses.

Simply put, as body size diverges among species (by adaptive evolution), individuals changing their size also must change their metabolic rate. If the change is the "wrong" amount (according to a particular model), the individuals are either no longer functional or have excess capacity and thus have evolved or retained an unnecessary inefficiency.[1]

From a causal point of view, then, in these models the exponent of metabolic scaling is a secondary result of a constraint that acts to maintain efficiently an invariant functional criterion at all sizes. This criterion has a geometrical basis, and so the widespread appearance in empirical studies of values clustering near the canonical values that the models predict has a strong nonadaptive component. In fact, one can describe this allometry as being largely the result of nonadaptive selection on the function and efficiency of transport networks or of bodies needing to dissipate heat. Thus, one would expect to see the allometry as a lawlike generalization.

Similarly, we have argued that it is the ecological requirement for the size invariance of the average per-generation production of successful offspring that ensures the quarter-power scaling of lifespan, which otherwise is not tightly constrained by physiology (see chapter 5). Further, we have seen that the scaling of animal abundance may be explained by a process that maintains an independence of population energy use and size (see section 2.3). In both of these cases, we have also been able to offer explanations of the allometries in terms of nonadaptive selection processes, and it is not surprising that in both of these latter cases a property that is invariant with size is the ingredient for a successful explanation. Other allometric relationships that are logically derivable from these two invariances (such as the scaling of number of heartbeats per lifetime from both physiology and life history or the scaling of population density from the scaling of energy use) are not necessarily anything more than reflections of what is causing or maintaining the invariances and do not necessarily require additional causes to explain them.

Thus, many allometric relationships in biology that at first seem to be independent and to require their own specific explanations may in fact belong to small networks of logically related allometries, whose members all descend functionally from one causal invariance. Such networks can be linked to other allometric networks when one of the network's allometries forms a component of a different allometric or functional invariance, subject to different causal mechanisms. In this way larger networks of coordinated scaling relationships can be formed that involve a wider array of biological properties and systems than is addressed by any of the subnetworks alone. A significant example may be the supernetwork involving metabolic scaling and ecological power laws that links physiology, life history, and ecology, which we discuss in section 6.2.

6.2 The Coordination of Ecological and Metabolic Power Laws

One thing that has impressed ecologists in recent years is the degree to which key ecological variables that are known to vary allometrically with body size—involving ecologically important quantities such as abundances, productivities, and demographic characteristics—seem to be interconnected in ways that ultimately appear tied to individual metabolic rates (Calder 1984; Brown et al. 2004). Brown and colleagues have elevated this and its biological implications to a general theory, which they call the metabolic theory of ecology (MTE) (Brown et al. 2004; Brown, Sibly, and Kodric-Brown 2012). According to the MTE, it is ultimately the scaling exponent for metabolic rate that determines the values of the other scaling exponents in this supernetwork, which tend to take on quarter-power values ($\pm 3/4$, $\pm 1/4$), or in the cases of both lifetime mass-specific metabolism and lifetime number of physiological cycles, an invariance (0). The fact that the MTE can predict at a general level a wide variety of scaling relationships does not mean that it currently provides an adequate causal explanation for why it does so. However, we assume that the MTE's general success in predicting empirically observed scaling relationships means that future research will elucidate why this is so and at what level of resolution it may fail (Duncan, Forsyth, and Hone 2007; Price et al. 2012; Glazier 2015).

Here we recognize that the MTE predicts rather well the observed exponents across the supernetwork, at a sufficiently general level of analysis. We would like to suggest, however, that its ability to do so depends on the existence of certain traits whose invariant relationships with size are responsible for the coordination of the exponents beyond physiology and across the supernetwork.

In figure 6.1 we have presented one view of part of this interrelated supernetwork of scaling relationships involving the three realms of physiology, life history, and ecology (cf. Maiorana 1990). In each balloon we have

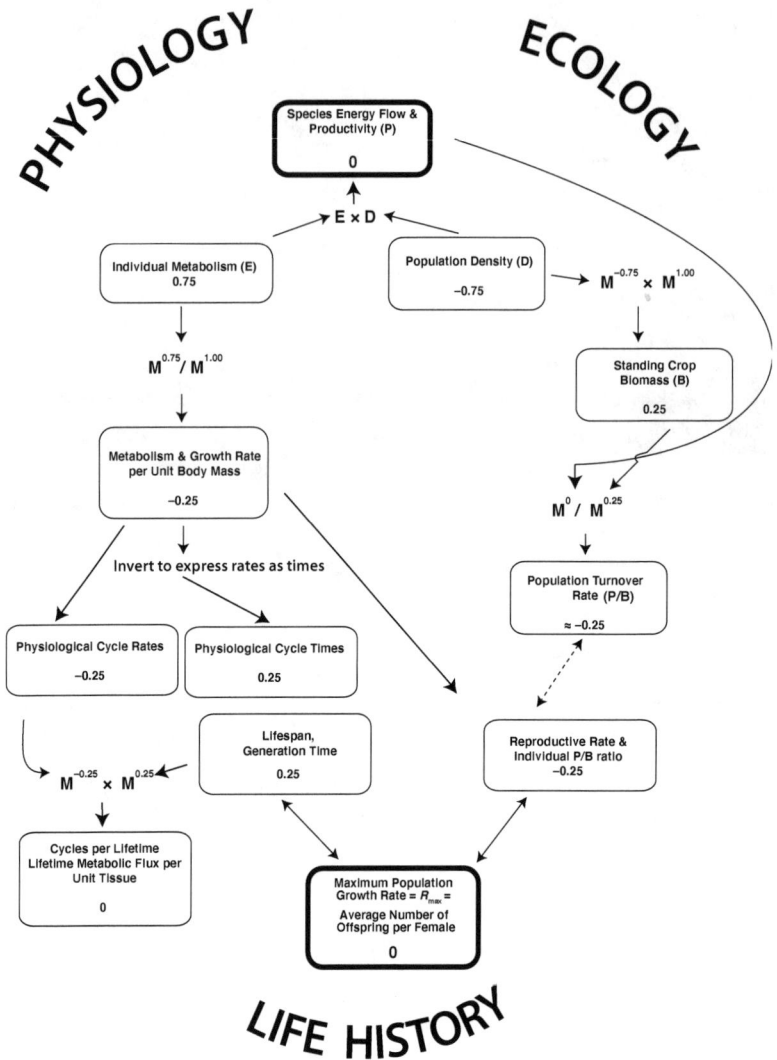

FIGURE 6.1. Supernetwork of physiological, life history, and ecological allometries. Balloons indicate the body-mass scaling exponent of the contained variable(s). Arrows indicate dependencies or links between variables and the algebraic basis of the transformations of the exponents, if applicable. For detailed explanation, see text. *Note: E* and *D* (*top*) must scale reciprocally if the energy-use invariance holds. Population turnover rate (production-to-biomass ratio) and individual reproductive rate and productivity (*bottom right*) must scale the same, by definition. Invariance of R_{max} (*bottom*) links individual reproductive rates with lifespan variables (Ginzburg, Burger, and Damuth, 2010); the invariance forces the connected rates and times to scale reciprocally.

listed a biological property (or set of properties) and a representation of the characteristic power by which that property scales with body mass. Actually, these are *ideal* values based upon simple transformations such as the ones described in the figure. Empirical observations are subject to uncertainty, sampling, and additional constraints not considered here; nevertheless, empirical observations agree well with the figure's values (Blueweiss et al. 1978; Peters 1983; Calder 1984; Morin and Dumont 1994; Niklas and Enquist 2001; Ernest et al. 2003; Savage et al. 2004; Atanasov 2005, 2007, 2012; Charnov, Warne, and Moses 2007; Ginzburg, Burger, and Damuth 2010; Brown, Hall, and Sibly 2018; and see Hatton et al. 2019).

In the figure, in going from one balloon to another, we have indicated how the source balloon's exponent combines with an algebraic constant or with the exponent of another balloon to yield the scaling of a new biological property. So, starting in the upper left-hand corner, we have individual metabolic rate and 0.75 — the canonical relationship for whole-body metabolism discussed earlier. (Remember, we are using this specific value for concrete illustration only and are making no claim here for its universality. The reason that we use a specific value will shortly become apparent.) Moving downward from metabolic rate, metabolic rate per unit body mass is simply the whole-body metabolic rate divided by body mass. (Body mass itself scales simply as itself, i.e., as mass (M) raised to the power of 1.00.) In performing a division, we subtract the denominator's exponent (1.00) from the numerator's exponent (0.75), which yields −0.25. In addition, we have reasoned that the scaling of growth rates should scale with the same exponent as per-unit metabolic rate (−0.25), since the rate at which cells divide and tissue grows should reflect energy flux at the cellular level; this is also empirically confirmed at the largest scales for both plants and animals (Damuth 2001; Niklas and Enquist 2001). The rest of the figure is made the same way, adding new properties and predicting their scaling exponents by simple algebra involving the exponents of their related properties.

Note that here we tend to see quarter-power exponents throughout, for relationships in all three realms. This appears to support the MTE in its original form. But the appearance of these quarter powers throughout the whole network is dependent upon two critical linkages.

At the top and bottom of the figure, respectively, we have highlighted two balloons in bold, where invariant relationships (exponents of 0) are located. At the top of the figure, the rate at which a species population in a trophic level uses energy per unit area (its energy flow and, by extension, its productivity) is the product of metabolism (scaling as mass to the 0.75 power) times the number of individuals (which is the population density, which has its own balloon and scales with mass as –0.75). In performing that multiplication, we add exponents, which in this case sum to 0, as depicted in the species energy flow and population productivity balloon. This invariance connects metabolic demands with population and community ecology (through energy control) and represents energy equivalence: Damuth's Law. At the bottom of the figure we have the invariance of R_{max} described in chapter 5, which is the result of linkage between individual growth and reproductive rates (scaling as mass to the –0.25 power) and average adult lifespan (which, to maintain the invariance, must scale as mass to the 0.25 power). This invariance thus connects life history ecology to reproductive physiology and determines lifespan (Ginzburg, Burger, and Damuth 2010).

In chapters 2 and 5 we have presented nonadaptive selective explanations for why we should routinely observe these two invariances, which represent lawlike generalizations that provide the crucial links among broad sets of algebraically related allometries. In each case, the quantity that is scaled against body mass is the combination of two other quantities that scale reciprocally with body mass (i.e., the sum of their exponents is 0). The two quantities that must scale reciprocally are indicated in the figure above or below each invariance balloon. Each of the two invariances indicates something that is functioning the same way, on average, at all sizes.

Note that there is a third invariance in the figure, the number of physiological cycles per lifetime (i.e., the number of repeated physiological events, such as heartbeats or breaths). We regard this invariance as a secondary consequence of the R_{max} invariance, because we think the latter more likely determines the lifespan. In chapter 5 we explained why we prefer this interpretation to that of the direct determination of lifespan by metabolic rates.

Now we can point out one surprising but simple attribute of this network. In general the values that the powers take on in the coordination network are determined by the specific value of the exponent of metabolism (0.75 in the figure). If one were to change *only* the exponent of metabolism, all of the other values would change, *except* for the two invariances caused by nonadaptive selection. The invariances are unchanged because of the reciprocities enforced in the case of the two highlighted invariance balloons. So, for example, if the metabolism exponent were made to be 0.67, every exponent that is ±0.75 would change to ±0.67, and every exponent that is ±0.25 would change to ±0.33. The invariances, because of the reciprocality constraints in each of the two highlighted invariances, do not change. At first sight, these invariances would appear to confirm that the exponent of metabolism "determines" the values in the network.

However, this is only because the nonadaptive invariances hold. If, instead of changing the metabolic exponent, we break these two laws and allow the exponents of population energy use and R_{max} to have nonzero values, we can see that metabolic rate would no longer exhibit a simple determination of scaling exponents throughout the network. Under the canonical metabolic rate exponent of 0.75, population density and standing crop biomass take on the familiar quarter-power exponents only if the energy flow invariance simultaneously holds. If we leave metabolic rate alone and break the invariance, allowing species energy flow to scale with an exponent of, say, 0.20, then population density would then have to scale as −0.55 (because $E \times D \propto 0.20$, so $D \propto 0.20 / E \propto 0.20 - 0.75 = -0.55$). In this case standing crop biomass must scale as 0.45, and both of these exponents are substantially different from the quarter-power values (−0.75, 0.25) seen when the invariance holds. Likewise, if we break the lifetime offspring invariance (R_{max}), this allows lifetime to scale as something other than a quarter-power exponent, in which case mass-specific lifetime metabolism or cycles per lifetime no longer show an exponent of 0.

So although the scaling of metabolic rate with body mass does in some sense determine the particular canonical values expected for the various exponents, the reason that the whole supernetwork hangs together appears to be the two invariances that we argue are maintained by different processes of nonadaptive selection, not by metabolic rate. Even if our mechanisms for the two causal invariances are wrong, we still think that those invariances are the points at which nonphysiological considerations enter the supernetwork and translate the effect of metabolic scaling from physiology to other realms. Ian Hatton et al. (2019) identified these two invariances as being particularly intriguing relationships but did not advance a selective explanation for their importance.

One could expand the figure to include a variety of other related characteristics within each realm, but we have chosen what are probably the most important ones for illustration. Adding additional balloons would increase complexity but would not markedly change the overall scheme (For an expanded version of the life history portion of the network, see Maiorana 1990, fig. 6.6). An extended focus on the energetic content of biomass and reproduction reaches compatible results (Brown et al. 2018). Extending beyond the depicted supernetwork should be possible but might require additional invariant linkages. For example, extending the population ecology relationships into the ecological use of space (e.g., home range) would require resolution of issues that are currently not well understood (Damuth 1981a; Haskell, Ritchie, and Olff 2002; Jetz et al. 2004).

In reality, for every characteristic there are likely to be influences from many sources that cause empirical deviations from the expected canonical exponents. Such deviations are real phenomena that may apply at some scale and derive from the fact that the relationships are among actual organisms and populations that are subject to myriad specific local constraints and causes (Glazier 2005, 2015). In contrast, our suggestion that the supernetwork requires invariant linkages that are not straightforward deductions from metabolic rate would be unchanged if global metabolic rate were altered, but then the expectation for everything else other than the invariances would also change.

Current theory (West, Brown, and Enquist 1997; Banavar et al. 2010) may be able to explain why the canonical value of metabolic rate is approximately 0.75, and quarter-power exponents follow for physiological variables. However, the two invariances that we have identified are the points at which ecological processes necessarily interact with the physiological network of exponents, and understanding how the invariances are maintained would complete the picture. Nonadaptive selection dynamics are the explanation in each case.

CHAPTER 7

Nonadaptive Selection in Multispecies Interaction Networks

7.1 Introduction: Observed, Lawlike Regularities in Multispecies Interactions Are the Result of Nonadaptive Selection

We have already briefly discussed two examples of nonadaptive selection in multispecies interaction networks: length of trophic chains (chapter 1) and selection on local stability of three-species modules that make up building blocks of a food web (chapter 2). The stability of communities and their components has been of long-standing interest in ecology, and the theoretical exploration of these topics has grown rapidly in recent years. Although seldom made explicit, much of this research directly implies that numerous nonadaptive selection processes are at work structuring biological systems at this level.

However, before we can discuss more generally nonadaptive selection in multispecies ecological systems, we need to focus on a series of issues involving the interactions between a single predator species and its prey. As we will see, nonadaptive selection processes ensure that individual predator-prey relationships tend strongly to exhibit a number of significant characteristics. The effects of these ramify upward, affecting both the evolution of predator-prey interactions over time and the existence of yet other nonadaptive processes at the level of multispecies systems.

7.2 Consumer Interference

7.2.1 Single-Species Population Dynamics

First let us consider the dynamics of populations of a single species. It is axiomatic that an ecological population would grow exponentially in the absence of limiting factors (Ginzburg 1986; Ginzburg and Colyvan 2004; Dodds 2009). The nature of these limiting factors was the topic of considerable debate in the mid-twentieth century (e.g., McLaren 1971). In particular, the question was: Are these factors density *independent* (in which conspecific individuals do not interfere with each other's growth and reproduction, but something else is responsible for changes in population size), or are they density *dependent* (in which interference among individuals for limiting resources provides negative feedback on population growth as density increases)? Resolution of this issue reveals a role for nonadaptive selection.

An equilibrium in a density-*independent* model is possible if the birth rate is exactly compensated for by the death rate. Such an equilibrium is, however, structurally unstable. That is, a small change in either the birth rate or the death rate will either (1) lead to extinction or (2) lead to explosive exponential growth, which will be stopped eventually by some limiting factor (food, space, light, etc.). The alternative is density *dependence* (interference for the limiting resources represented by a set of specific limiting factors). Herbert Andrewartha and L. Charles Birch (1954) advanced a strong density-independent view of natural populations (random walks of abundance, away from resource limitations). But such populations would not exhibit a stable equilibrium and would eventually explode or die. The density-independent view has generally lost out among ecologists, and today we think that structurally stable equilibria, involving density-dependent mechanisms, are the only ones to be seen (Begon and Townsend 2021). Structurally unstable populations, without density dependence, have been selectively removed from our view. This is, in fact, a straightforward example of a nonadaptive selection mechanism, although it is not usually explicitly recognized as such.

One of the most common and general mechanisms of density-dependent regulation must be as follows: as the population grows, the competition among individuals for resources through direct or indirect interference increases as the per capita availability of those resources decreases. Consumer populations of all kinds are subject to the same considerations, so it stands to reason that when the resource is another species, the predators in predator-prey models should often exhibit some level of interference among their members. The next two sections show how nonadaptive selection based on levels of mutual interference among predators affects the observed dynamics of predator-prey interaction systems.

7.2.2 Predator-Prey Models: Arditi-Ginzburg Conjecture

Standard ecological predator-prey models take the form of a pair of population growth equations, one for the predator and one for the prey. Linking these equations are interaction terms that represent the response of each population to the presence and behavior of the other—ultimately defining the rate at which the predator population consumes prey and the rate at which the predator population grows as a result of that consumption (these are called the *functional response* and the *numerical response*, respectively). There are two ideal versions of these response terms as implemented in models, which have been termed *prey-dependent* and *ratio-dependent* (Arditi and Ginzburg 1989). Traditional predator-prey models incorporate purely prey-dependent responses, which base the growth rate of the predator population entirely on the abundance of the prey; the size of the predator population in relation to the prey has no direct effect upon the predator's consumption rate. In ratio-dependent models, in contrast, the growth rate of the predator population depends on the ratio between predator and prey population sizes (i.e., the amount of prey available to each predator, on average). In the ratio-dependent case, increasing predator abundance relative to that of the prey depresses the average predator's food intake.

The conversation in the literature has heretofore mostly focused on a distinction between these two different theoretical starting points, prey dependent and ratio-dependent. It is useful to think of these two as making different assumptions about the degree of mutual interference among predators.[1] Prey-dependent models assume that there is no interference.[2] Ratio dependence describes a situation in which there is, on average, equitable sharing of the prey resource. These two versions of predator-prey interaction refer to different points on a continuum, based on the degree to which the predators are interfering with each other's consumption. When predator population densities are very low, the predators are unaware of each other, and their rate of consumption depends mostly on the rate at which they encounter and process prey; interference is essentially zero.

Prey dependence will describe the population well in this situation. But as predators become more abundant, they will begin to interfere with each other. They will consume prey that another individual predator might have encountered, and so on, and under these conditions the key issue for the growth of the predator population is the amount of prey that is available to each predator, given the prey and predator abundances. Thus a model in which per capita prey availability is the key—ratio dependence—describes the system better when the predator density is high. Roger Arditi and Lev Ginzburg (2012) thus proposed that the most general type of model (one making the fewest explicit biological assumptions) describing a simple predator-prey system over time would be one they termed *gradual interference*, in which the interaction gradually changes from one of prey dependence to one of greater and greater degrees of interference as the relative density of predators increases toward equilibrium (Abrams and Ginzburg 2000). The same kind of reasoning and same kind of model can apply to different predator-prey pairs that exhibit different levels of interference. Such *consumer-dependent* (or *predator-dependent*) models add a single additional parameter, m, to the ratio-dependent model that describes the level of mutual interference among the predators. These models can be thought of as ones in which consumer population growth depends on both the numbers of prey and the numbers of consumers, to varying degrees (Hassell and Varley 1969; Arditi and Akçakaya 1990; Arditi and Ginzburg 2012). The value of m is never negative. When $m = 0$, these models reduce to the prey-dependent models, and when $m = 1$ they become identical to ratio-dependent models.

The preceding reasoning suggests that, in turn, if we were to look at values of consumer interference, m, in natural populations, they would be spread across a range of values ≥ 0, and the consumer-dependent models would be good descriptors of the populations' dynamics. In fact this turns out to be the case. Figure 7.1 shows the frequency distribution of empirical m values in a large sample of natural and naturalistic experimental populations compiled by Marc Novak and Daniel Stouffer (2021). The frequency distribution shows not a random dispersion throughout the continuum, but rather one with a distinct peak.

For historical reasons, predator-prey systems have been modeled almost exclusively as if populations were always at the extreme prey-dependent end of the spectrum (corresponding to $m = 0$). This extreme seems to be rare or absent in nature. It is clear that most populations actually experience higher levels of interference, and the modal value is in the

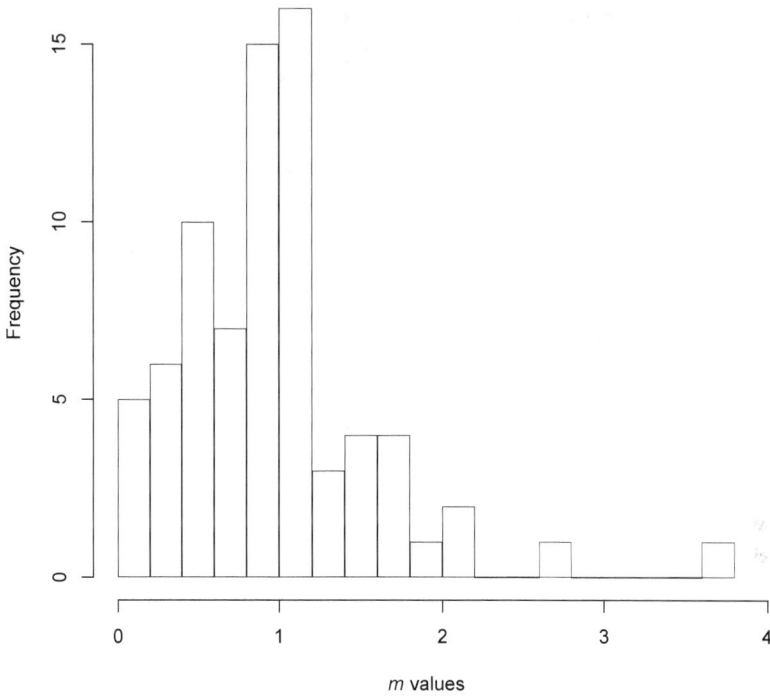

FIGURE 7.1. Frequency distribution of the values of the consumer interference parameter m estimated from the literature by Novak and Stouffer (2021), assuming the model of Arditi and Akçakaya (1990). The 75 cases range in value from 0.034 to 3.79 (2 cases for which authors were unable to obtain reasonable standard errors are omitted). The truncation at 0, positive skew, and a modal value near 1 are evident. Figure from Ginzburg and Damuth (2022).

neighborhood of pure ratio dependence ($m = 1$). This appears increasingly to be the accepted view (Skalski and Gilliam 2001; DeLong and Vasseur 2011; Arditi and Ginzburg 2012; Molles 2019; Novak and Stouffer 2021).

As previously argued, fairly high levels of interference should be expected in most natural populations. The empirical, peaked, somewhat positively skewed distribution of values suggests processes acting to remove or limit both very low and very high values of interaction. We argue in the next section that nonadaptive selection processes guarantee that most of the time when we observe predator-prey systems in nature, they will be clustered around an intermediate value of m close to 1. A theoretical article by Yuri Tyutyunov, Deeptajyoti Sen, and Malay Banerjee (2024) based on a very different dynamical model reaches a similar conclusion:

stable outcomes are concentrated at intermediate levels of interference, with both extremes tending to be unstable.

7.2.3 Structural Instability at the Prey-Dependent End of the Spectrum, the "Paradox of Enrichment," and Overcompensation

At the prey-dependent extreme, where predators do not interact, an equilibrium is quite possible; it can even be stable (as in a Lotka-Volterra cycle) with respect to initial conditions (Lyapunov stability). However, this equilibrium is structurally unstable; a number of likely ecological and evolutionary changes in the system, including improvement in predator or prey efficiency, enrichment of prey resources, and increase in prey reproductive rate, can all lead to instabilities, followed by extinction of one or both species. So, in the prey-dependent range of the gradual interference continuum, extinction is highly likely. However, with increasing mutual interference among predators, the equilibrium becomes progressively more stable. In fact, it is structurally stable when $m = 1$. Stability of the equilibrium where interference is high changes little if parameters of either interacting species evolve in any direction (Arditi and Ginzburg 2012).

One of these causes of instability at the prey-dependent end of the continuum is well understood and is widely known as the "paradox of enrichment" (Rosenzweig 1971). Imagine that we have a stable predator-prey system in the prey-dependent region, at or near $m = 0$. Now, imagine that conditions improve for the prey, such that the carrying capacity of the prey increases. Intuitively, things should be better for both predator and prey. However, the prey-dependent models show that enrichment of the environment for the prey can actually *destabilize* the system, and if prey carrying capacity continues to rise, either the predator or both species will go extinct. It is paradoxical that an apparent benefit destabilizes the system and causes its demise. This *paradox of enrichment* has become an axiom in ecology, although it is seldom, if ever, observed in nature (Jensen and Ginzburg 2005; Arditi and Ginzburg 2012). The paradox of enrichment describes an instability that occurs when interference in predator-prey systems is low or absent, near the prey-dependent end of the spectrum; the prey-dependent model is not a common natural phenomenon but the (correct) mathematical consequence of what appears to be an inappropriate model for predator-prey systems in general. All the empirical data that we know point toward strong interference being the rule. This is not because equilibria with lower levels of predator interference never existed, but because they were structurally unstable if they did.[3]

Thus, this is a typical process of nonadaptive selection. What happens with this interacting pair of species is analogous to the situation among populations of a single species, in which nonadaptive selection removes structurally unstable populations. Similarly, we are not likely to see extant pairs of predators and prey close to the prey-dependent limit, even though it is logically possible, and even though such pairs might be assembled in laboratory experiments (Mittelbach and McGill 2019, ch. 11). Nonadaptive selection based on structural instability will tend to eliminate these cases in nature, and only those with a degree of interference among predators will survive.

However, interference cannot be too strong, or new problems will arise. Ratio dependence ($m = 1$) corresponds to equal (on average) access to prey for all predators; that is, access to prey depends only on how many prey there are per predator. However, we know that sometimes predators exhibit higher levels of predator interference ($m > 1$; see section 7.2.2). Various biological mechanisms could lead to this "overcompensation." Predators might increase their antagonism with one another faster than the prey availability increases, with the predators wasting more energy fighting than they can gain by greater access to the resource. Or prey escape or avoidance behavior might escalate with increasing predator density, leading to the same effect. Whatever the biological details, the consequences for the predator population are severe; all other things being equal, as m exceeds 1 equilibrium predator abundances decrease.

At $m = 1$ (pure ratio dependence), there is an equitable distribution of the prey to each predator. As m exceeds 1, the predator population starts experiencing a net loss in prey resources, because their mutual interference is so great that it prevents that equitable distribution; predators "waste" time and effort in interference and capture less food per capita. The result is that above $m = 1$, we can have a stable equilibrium (in the Lyapunov sense), but the number of predators (P) will decline strongly with m. The rapidly decreasing equilibrium predator population sizes should be associated with greater risk of extinction in the long run, as a result of chance perturbations and other events.[4] Thus, this difficulty of keeping predator numbers high enough to avoid random extinction makes it unlikely that predator-prey systems will persist at very high m values (consistent with the conclusions of Arditi et al. 2004).

This mechanism now creates a *second* nonadaptive selection process, this time selecting against values of m that are too *high*. The predicted pattern that results from the *joint* action of these two different nonadaptive selection processes would be exactly what we see empirically (and globally) in figure 7.1. In this case, the two nonadaptive selection processes together result in an effect similar to stabilizing selection, maintaining the predominance of intermediate, highly stable levels of predator interference that are near ratio dependence. This case is similar to the lawlike energy equivalence relationship (discussed in chapter 2) in which nonadaptive selection acts like a regulator. The variance about the modal value here reflects real biological variation rather than just statistical observation error, so a wide range of values is feasible. However, the further the predator-prey system deviates from the most stable state, the less likely it is to survive and be observed.

So the most stable configuration should be ratio dependence, yet the empirical distribution is skewed such that the modal value of m is somewhat lower (perhaps near 0.90). Considered from a dynamic, nonadaptive selection perspective, this asymmetry around ratio dependence in figure 7.1 may simply be due to the two nonadaptive selective processes typically being of different strengths. In other words, the reason that the modal value of m shows a value less than 1 may be that, as m changes, populations lose stability at a faster rate to the right of 1 than they do to the left of 1. Thus it is not necessary to postulate any biological significance for a particular observed modal value of this distribution (such as 0.90). The observed value may not derive from a specific regularity of predator-prey dynamics, but rather may just reflect the asymmetry of the two nonadaptive selection processes just described. We expect that future research will clarify these issues (Novak and Stouffer 2021), but now, at least, it appears clear that equilibrium levels of predator interference that characterize the prey-dependent extreme are rare in nature.

The literature has often portrayed this topic in terms of a stark choice between two theoretical options for building models of predator-prey dynamics: pure prey dependence and pure ratio dependence. This treats them as if they were two direct, alternative interpretations of "true" predator-prey dynamics. A more nuanced view (consumer dependence) recognizes that populations at equilibrium will be spread out across a continuum of interference values (e.g., m). Nevertheless, such a view has tended to portray the continuum as running from prey dependence ($m = 0$) to ratio dependence ($m = 1$); that is, the two values are each extremes at opposite ends of the continuum. In fact, empirical studies show that values of m often exceed the value of 1. (The lower limit is still 0.) In contrast, a stabilizing nonadaptive selection view sees pure ratio dependence ($m = 1$) not as an end point, but rather as an intermediate value that is associated with maximum stability, in the sense of resistance to extinction.

These evolving views lead to some significant general conclusions. First, if one knows nothing about the level of interference among consumers, ratio dependence seems to capture the essence of the typical cases quite well, but prey dependence does not. Thus, ratio dependence, while not doing full justice to the actual variation in natural populations, is in this situation "less wrong" than prey dependence (Abrams and Ginzburg 2000, 340–41; Ginzburg and Damuth 2022). As such it should provide a better starting point for thinking about predator-prey interactions. Second, when researchers want to include realistic predator-prey behavior in theoretical models or in simulations, they have historically almost always started with a purely prey-dependent model—Lotka-Volterra or one of its descendants. Given both theory and empirical observations, doing this without substantial justification no longer seems to be the best practice.

7.2.4 Asymmetry of Interspecific Interactions

There is one more issue that we need to cover before we address multispecies systems and ecological communities. The mutual interactions *within* a species (say, a predator) that we have been discussing will be *symmetric*; on average the strength of each individual's effect on its neighbor will be the same as its neighbor's effect on it. Once we switch to the study of communities and their multispecies components, we include species interacting with each other. In section 2.2 we briefly introduced the community interaction matrix and quasi-sign stability (QSS), which form the basis of our discussion of community patterns. (Interactions among species are depicted there as the off-diagonal elements a_{ij} of an interaction matrix, where $i \neq j$.) Current perspectives and observations suggest that most of the time interspecific interactions will be *asymmetric*, as detailed in this section.

Beginning with Lotka (1925) and Volterra (1926), and continuing through May (1973a), species interactions were almost always treated in theoretical work as being symmetric; that is, the (a_{ij}, a_{ji}) pair of interactions between a given two species had values of equal or similar magnitude. In responding to May's 1973 work, researchers at first continued to restrict themselves to the assumption of symmetry and the exclusive reliance on Lotka-Volterra equations (Roberts 1974; Tregonning and Roberts 1978, 1979; Sugihara 2017 [originally written in 1983]). The major assumptions were either symmetry ($a_{ij} = a_{ji}$, for competition) or precise antisymmetry ($a_{ij} = -a_{ji}$, for predation). In some cases *all* of the interactions among all of the species were set to an identical value.

However promising, this early work involving the stability properties of model communities with symmetric interactions did not continue to generate new insights. Under symmetry, theory led to the conclusion that stability of communities would occur only if interactions were weak (or absent). This is a well-known result in ecology. However, in nature we see large and diverse communities with at least some strong interactions. This might have led to reassessment of the assumptions of the theory, but instead it left ecologists with the persistent impression that theory and observation in this area of biology were inevitably in conflict.[5]

However, there never has been any biological justification for interaction symmetry, nor any biological explanation of why it should be universal and precisely symmetric. In fact, interaction *asymmetry* has long been observed in natural communities (Schoener 1983; Persson 1985; Jordano 1987). Current views reinforce the observation of widespread asymmetry of species interactions (see section 7.4.4).

Interaction asymmetry derives from a wide array of sources, including aspects of predator-prey dynamics ("donor control" in Arditi and Ginzburg [2012, 22–24, 80–82]),[6] effects of size and size-related traits (see section 7.5), differences in function between types of species (e.g., bipartite networks; see section 10.12.3), and innumerable other biological circumstances. Multiple influences can be active simultaneously, and which factors determine the actual level (and polarity) of asymmetry in a given case depend on biological details of the interactions. Some asymmetry is the norm.

With this perspective, we can now move forward with our discussion of nonadaptive selection among the components of multispecies systems.

7.3 How Nonadaptive Selection Explains Short Trophic Chains

In section 1.2 we briefly described a widely known but puzzling ecological observation, that trophic chains are usually quite short. Here we expand upon the topic in order to describe more fully how we think nonadaptive selection explains this pattern.

At first sight short trophic chains imply an unresolvable dilemma: we saw in chapter 2 that a community module that is a three-level trophic chain is qualitatively stable and widely observed; *any* values of the elements of the matrix representing interactions in such a structure are stable, as long as the signs of the interactions remain the same. In fact, pure trophic chains like this are qualitatively stable structures *regardless* of their length (May 1973b). So why do we not see ecosystems with seven, or eight, or any large number of trophic levels, with many long trophic chains? In nature, communities are usually composed of no more than three or four levels, and, more counterintuitively, the number of levels does *not* correlate with primary production. Very small ecosystems comprising few species and that are subject to high levels of disturbance may be likely to have chains of limited duration (Guo et al. 2023), but that does not explain why large ecosystems are similarly constrained. There are three to four trophic levels in the tropics, the same number as in the Arctic tundra. There may be very different numbers of species per level but no difference in the number of levels (Cohen, Briand, and Newman 1990; Pimm 1991). Analogs of the highly stable trophic chain are seen in nonecological networks as well. We think that the explanation for limited trophic chain length in biological communities lies in a specific property of organisms, that is, the difficulty of ensuring an entirely *pure* trophic chain, in which each species consumes *exclusively* the next species lower in the chain. In reality many chains exhibit a certain degree of omnivory—species consume resources from more than one level in the chain. If species 1 eats 2, and 2 eats 3, species 1 may prefer species 2, but may also obtain benefit from eating species 3 if opportunity permits. Since predators are in general much larger in body size than their prey (see section 7.5), top predators usually have little difficulty capturing and consuming resources from multiple, or even all, lower levels. Furthermore, it is possible that a species lower down the chain may have the ability to feed "upward" as well; say, species 3 has a way to extract energy from species 1 but is still subject to predation from species 2. Thus a realistic view of trophic chains should allow for some degree of omnivory as a common occurrence.

A perfect trophic chain for four species looks like figure 7.2, in which species 1 eats species 2, and so on.

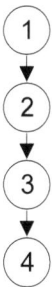

FIGURE 7.2. Perfect trophic chain for four species, each arrow extending from a predator to its prey

Its corresponding matrix is

$$
\begin{bmatrix}
- & + & 0 & 0 \\
- & - & + & 0 \\
0 & - & - & + \\
0 & 0 & - & -
\end{bmatrix}.
$$

and this is qualitatively stable.[7] The positive effects of each prey species on its predator are above the diagonal, and the negative effects of each predator on its adjacent (and only) prey are below the diagonal. In contrast, a four-species trophic chain with complete omnivory (where each species feeds on all below it in the chain) looks like figure 7.3.

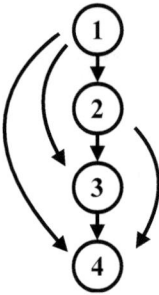

FIGURE 7.3. Four-species trophic chain with complete omnivory

The interaction matrix for such a system would look like

$$
\begin{bmatrix}
- & + & + & + \\
- & - & + & + \\
- & - & - & + \\
- & - & - & -
\end{bmatrix}.
$$

This matrix is much less stable, judging from a QSS analysis of the kind we described in section 2.2. The 0s in the stable matrix have been replaced by the pluses and minuses of the omnivory links, and this has contributed to instability (see appendix 10.12 for more details).

Recall that asymmetric interactions are common (see section 7.2.4). Thus, in the preceding matrix the negative values on one side of the diagonal might be expected to be different in magnitude than the positive values on the other side of the diagonal showing the prey. This asymme-

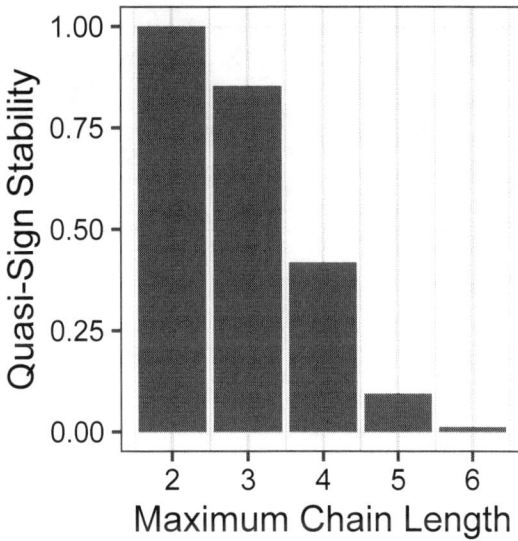

FIGURE 7.4. QSS of omnivorous food chains of different lengths (Borrelli and Ginzburg 2014)

try *does* improve the stability of an omnivorous trophic chain of a given length (see section 7.4), but it does not change the pattern of relative stability across chains of different lengths. Progressively longer chain lengths are less and less stable, based on the QSS analysis by Jonathan Borrelli and Lev Ginzburg (2014) shown in figure 7.4. The results shown are for simple, fully omnivorous trophic chains with moderate predator-prey interaction asymmetry. However, Borrelli and Ginzburg also found the same qualitative pattern among trophic chains comprising a variety of randomly generated matrices constructed with different numbers of connections among the species, as well as other structural characteristics. Chains of three, or four, or perhaps even five levels are viable, but longer chains are very unlikely (see figure 7.4).

So we think that short trophic chains, limited numbers of trophic levels, and independence of both from primary productivity are explained by nonadaptive selection on the stability of imperfect trophic chains. When the general source of variation in stability is recognized, the explanation appears to be a simple macroscopic consequence of nonadaptive selection.

In fact, one can wonder why it is that we believe communities can be characterized by "trophic levels" in the first place. The number of trophic levels we recognize must roughly reflect the average lengths of trophic chains. But the very concept of a trophic level can be a substantially approximate abstraction (Thompson et al. 2007; Ginzburg and D'Andrea 2023). The more similar the omnivorous web is to a chain (with weaker interactions two or more steps away from the pure chain), the closer it is to a qualitatively stable structure (May 1973b; see also appendix 10.2). If, due to this stability criterion, the omnivorous structures similar to pure chains have been retained by nonadaptive selection, then this would account for what we see today: arrays of similar, *approximately* hierarchical chains of a limited number of links that we perceive as together representing more or less distinct levels. We would not find much interest in trophic levels if there were a potentially unlimited number of them (pure chains) whose length varied at random among communities, or if there were few stable chains at all (strong universal omnivory). Trophic chains that are just imperfect enough to be limited in length, but not so perfect as to exclude omnivory, might be expected to dominate communities. So the perception of trophic levels, and their general applicability and utility as a concept, may largely be a consequence of the nonadaptive selection processes shaping communities.

7.4 Qualitatively Stable Structures, Triangular Matrices, and Asymmetry of Interaction Strengths

Our discussions of consumer-resource community modules (chapter 2) and trophic chains (this chapter) have highlighted a number of ways that nonadaptive selection may explain widespread, common structures in community networks. We have seen that very general structural properties of consumer-resource networks and their components—sign structure and asymmetry of interaction values—give rise to matrices with different relative stabilities, and selection on such stability favors the persistence of some communities (or their components) over others. That is, nonadaptive selection can sift among communities of different stability and may be able to explain why some of the structural features observed in natural communities are so widespread. Furthermore, we have seen that matrix structures that approach, but do not necessarily achieve, full *qualitative* stability can nevertheless still exhibit a significant degree of stability (QSS; see section 2.2 and appendix 10.2) and can thus be favored by nonadaptive selection, relative to less stable structures.

There is a growing body of research suggesting that certain structural motifs tend to characterize community networks.[8] These include different types of "hierarchical" structures, modularity, and the asymmetry of interaction strengths. We argue that these common motifs are just what we would expect to see if nonadaptive selection is generally operating on the relative stabilities of communities.

This view does not replace traditional approaches to community stability that aim to derive the quantitative conditions (in terms of interaction coefficients and population dynamics) for stability and coexistence, which are comparatively difficult to generalize. Rather, we take a qualitative approach that explains why we so often see what we see, regardless of the details of the processes by which individual communities have been built.

7.4.1 Stability Selection in Community Ecology and Traditional "Quantitative" Stability

Stability of populations and communities—and thus the possibility of selective processes winnowing what we ultimately observe—is a venerable idea in ecology. The foundational work of May (1973a) on model ecosystems ignited a renewed interest in these issues. Stability selection is an important part of many theoretical investigations involving population dynamics of interacting species and community structure. In many cases, selective explanations for community properties involving community stability are clearly nonadaptive, but may not often be explicitly identified as such (e.g., Mittelbach and McGill 2019; Amarakesare 2022) (See also Borrelli et al. 2015 for additional discussion).

The question that community ecologists traditionally ask of population dynamic models is whether there is a stable equilibrium in which the populations of all species are positive, unchanging, and robust to small perturbations. In order for such an equilibrium to be stable, the matrix of ecosystem interactions has to satisfy certain mathematical conditions, specific to the particular dynamic equations under study. Only by first specifying these equations can the question of the existence of a positive equilibrium be raised. We do not believe, however, that there is currently a single set of general ecosystem dynamics equations that can be relied on for this purpose. Nevertheless, considerable insight has been gained about processes that might confer stability under specific circumstances (see, e.g., Mittelbach and McGill 2019).

We might call this kind of stability *quantitative* stability, because it depends on the particular values of the interaction coefficients a_{ij}. If the matrix is not stable, the equilibrium is unstable, and we will not observe that community in nature (except, perhaps, temporarily). A large body of work has been devoted to finding out how often such quantitative stability is to be expected in randomly generated communities and what biological characteristics contribute to the stable equilibria (May 1973a). Quantitative stability (simply "stability" in most of the literature) certainly occurs in nature. Measured interaction intensities in some real ecosystems have been shown to form a matrix of interactions that is in fact stable (Neutel, Heesterbeek, and De Ruiter 2002).

However, quantitative stability is still subject to destabilization in the face of environmental variability or of evolutionary change. In both cases the intensity or sign of some of the interactions will change, and a community that was stable before may become unstable (Bolnick et al. 2011). Therefore, quantitatively stable ecological equilibria may nevertheless be fleeting in the long run: present, then absent, then present again. Yet core structural elements of diverse natural communities tend to be resilient and recognizable over long spans of historical and even geologic time. Such stability implies that our picture of community stability is incomplete.

7.4.2 Qualitative Stability

Qualitative stability offers a different way to look at the distribution of community structural characteristics. Let us consider some qualitatively stable matrices. We have already seen that some three-species community modules and all pure trophic chains are qualitatively stable (see sections 2.2 and 7.2). But the simplest and most general example of a qualitatively stable matrix is a *diagonal* matrix, with negative entries on the diagonal and 0s everywhere else:

$$\begin{bmatrix} - & 0 & 0 & \cdots & 0 \\ 0 & - & 0 & \cdots & 0 \\ 0 & 0 & - & \cdots & 0 \\ \vdots & \vdots & \vdots & \ddots & \vdots \\ 0 & 0 & 0 & \cdots & - \end{bmatrix}.$$

Ecologically, this interaction matrix says that each species in the community network experiences a negative feedback from its own abundance to its growth rate (which is just typical density-dependent population growth), represented by the diagonal entries. Otherwise, the 0s indicate that species do not interact in any way whatsoever. Every species has its own independent dynamics; how much more stable could a community be?

$$
\begin{bmatrix}
- & & & & \\
0 & - & & \text{Any values of} & \\
0 & 0 & - & \text{any sign} & \\
0 & 0 & 0 & - & \\
0 & 0 & 0 & 0 & -
\end{bmatrix}
$$

FIGURE 7.5. Structure of triangular matrix

Note that if we replace 0s with small numbers of any sign, the matrix will largely retain stability. So, it is reasonable to think of the diagonal matrix as a center of concentration of similar matrices, with interactions being weak in relation to self-interaction (the diagonal elements). How small is small? There is no general answer to this question, but May (1973a) addressed the issue and showed that the maximum magnitude of interaction values has to become smaller as the matrix becomes larger. This is a well-known generalization in ecology: it is harder to keep more species together if they interact, and even harder as the interactions become stronger and more numerous.

This generalization has long been regarded as implying either that communities in nature should not include many species interactions, or that the interactions that exist should be very weak; however, in contrast, we generally observe stable, diverse communities within which there are some strong interactions.

Let us consider a mathematically minor but biologically significant modification of the diagonal matrix. This is the *triangular* matrix, which, as in the diagonal case, has negative entries on the diagonal, but now *arbitrary* values on one side of the diagonal and all 0s on the other side (see figure 7.5).

This matrix, once again, is qualitatively stable. Interactions above the diagonal can be of any value, and stability is still assured. Significantly, small values instead of 0s below the diagonal will still yield some stability. How small are the values? How much stability? Again, these are not easy questions to answer with precision. But certainly, "as small is possible" will yield the most stable matrix. So from this perspective, a stable configuration would likely result from a restriction of most strong interactions to a triangle on one side of the diagonal, combined with very small or 0 values on the other side.[9] Is this combination biologically reasonable? And do we see it in nature?

7.4.3. Hierarchical Structures, Transitivity, and Asymmetry

How does the triangular matrix relate to the structures and topologies that students of community networks frequently observe? First, we draw attention to one of the properties of a community matrix. As we have seen, it is a square ($n \times n$) matrix, with the interactions indexed by pairs of numbers representing rows and columns, respectively:

$$\begin{bmatrix} a_{11} & a_{12} & a_{13} & a_{14} \\ a_{21} & a_{22} & a_{23} & a_{24} \\ a_{31} & a_{32} & a_{33} & a_{34} \\ a_{41} & a_{42} & a_{43} & a_{44} \end{bmatrix}.$$

If we number the species from 1 to n, then each element a_{ij} represents an interaction between two species. The point is that off-diagonal elements show a pattern: a given species pair is represented twice in the matrix, once on each side of the diagonal; the two interactions represent the reciprocal effects of the two species on each other. For example, a_{32} represents the effect of species 2 on species 3, and a_{23} the effect of species 3 on species 2, which could be different in magnitude, in sign, or both. Above the diagonal, effects are flowing from higher-numbered species to lower-numbered ones and below the diagonal from lower-numbered species to higher.

Now, the way we assign numbers to species is arbitrary. As long as the actual interactions among the species of each interacting pair are correctly depicted, community matrices with differently numbered species will have the same stability properties. But under different numbering schemes, a given species interaction may appear in different parts of the matrix. The matrix still has the same properties; what is true according to one arbitrary arrangement of the data is still true of all arbitrary arrangements of the data (see appendix 10.12 for examples).

What this allows us to do is to number the species strategically, in such a way that certain biological properties of the network are made manifest. For example, in section 7.3, when we constructed a matrix for a trophic chain with complete omnivory, we numbered the species in sequence from top predator to bottom of the chain and got the following:

$$\begin{bmatrix} - & + & + & + \\ - & - & + & + \\ - & - & - & + \\ - & - & - & - \end{bmatrix}.$$

All of the positive interactions (the effects of prey populations on their predators) fill the upper triangle, and the negative effects of the predator populations on their prey make up the bottom triangle. The structure depicted here is often described as a *transitive hierarchy*—transitive because each member of the hierarchy feeds on *all* of the species below it, and no species feeds "upward." As we remarked in section 7.2.4, values on one side of the diagonal are likely to be much smaller than the values on the other side. If the small values are small enough, then the matrix begins to resemble the triangular matrix. Thus, a community may approach qualitative stability if it can be arranged into a roughly triangular, hierarchical structure and, for whatever biological reason, the values of the species interactions are highly *asymmetric*—the (a_{ij}, a_{ji}) pair of interactions between two species typically have very different absolute values.

In many cases (such as the trophic chain we have been discussing) the asymmetry arises because of a particular biological property or constraint that differs with the *direction* of the interaction, which in turn refers to a side of the diagonal. Because the asymmetrical values of any species pair are found on opposite sides of the diagonal, it is often easy to arrange (number) species in such a way that this directional division is apparent. However, the key thing contributing to stability is that strong interactions are paired with weak ones (or zero values), and/or positive interactions are paired with negative ones. Both will tend to contribute *toward* a resulting qualitatively stable structure robust to most small environmental and evolutionary changes, though few actual communities are likely to be fully qualitatively stable.

If, on the other hand, all interactions were symmetric, long-term stability would happen *only* with *very weak* interactions *throughout*. That is indeed a generalization widely held in ecology, but it is not likely to represent fully the interaction structure of most communities that we see in nature. Although it is difficult to measure accurately interaction coefficients in natural communities, ecologists now have a considerable amount of information about them from indirect indicators of their absolute or relative strengths. The picture is in general not what we would expect from a universe of symmetric interactions among species.

7.4.4 Empirical Patterns in Community Interaction Structure: Hierarchies and Asymmetry

Up to now we have been discussing community matrices that represent food webs, but community ecologists frequently study other types of networks and sometimes depict them, and describe them, differently from the way we have been doing. One popular depiction is a *competition* matrix, particularly meaningful for situations in which the resource that supports the suite of species is not modeled as a population of organisms in the network, but as an external supply (such as sunlight or water). Another representation is the *bipartite* network diagram, in which the community is divided into two types of species that directly interact only with members of the other type and do not interact directly among themselves (e.g., mutualistic networks such as plants and pollinators; Bascompte and Jordano 2007). Food webs, competition matrices, and bipartite diagrams can depict most of the (usually partial) community networks of interest to ecologists, though combining them in theoretical work without losing information is still a challenge (Mittelbach and McGill 2019). All of these depictions, however, can be translated into an equivalent community matrix (see appendix 10.12).

So what do we find when looking at communities of these types? A growing number of studies have sought empirical patterns shared widely by natural communities, regardless of the degree to which those patterns may have been prefigured by current quantitative theory (Bascompte 2009). As a result, there is a developing consensus that natural communities are generally characterized by high levels of hierarchical structure and of interaction asymmetry. We have seen in previous sections how food chains often directly instantiate hierarchical structures and asymmetry; food webs, not surprisingly, have also been found to be built with "nested" subwebs that resemble the nestedness of mutualistic networks (Kondoh, Kato, and Sakato 2010). Likewise, distinct, highly transitive hierarchies associated with interaction asymmetry are observed in plant competition communities (Aarssen 1988; Keddy and Shipley 1989; Shipley 1993; Keddy 2001; Kinlock 2019) as well in communities of competing insects (Cerdá, Arnan, and Retana 2013), mammals (Hallett 1991; French and Smith 2005), and some marine invertebrates (Barnes 2002). Moreover, mutualistic networks generally exhibit significant levels of "nestedness" (which is equivalent to the hierarchical structures we have been discussing; see appendix 10.12), as well as associated interaction asymmetries (see Bascompte et al. 2003; Bascompte, Jordano, and Olesen 2006; Neutel and Thorne 2018).

Why is there a connection between hierarchical structure and interaction asymmetry? Hierarchies imply a structure involving one direction through the network. Arranging species interactions so that they appear as a hierarchy on one side of the diagonal involves ranking the species according to some biological criterion. For food webs, the hierarchical criterion is obvious; there tends to be a starting and ending position for a direction through the chain. For competition communities, the criterion consists of the relative pairwise competitive abilities of the species, with some species able to outcompete most others and some able to outcompete few; powerful competitors may exert strong effects on weaker species, who in turn exert little effect back on the winner. Interaction asymmetry is largely what allows this criterion ranking and thus what allows one to make the hierarchy to begin with (Keddy 2001). For bipartite mutualism communities, the criterion represents degrees of generalization or specialization—some species of each type interact with many species, others with few, and this creates a basis for interaction asymmetry (Jordano 1987; Bascompte et al. 2003; Bascompte, Jordano, and Olesen 2006). In all these cases, we can recognize biological reasons for how a given hierarchy criterion is related to the asymmetry of interaction values. In fact, because generalists and specialists are usually found not living alone, but in communities containing both, these stable community structures may help to resolve the well-known general problem of explaining the evolution of generalists and specialists (Futuyma and Moreno 1988).

Although hierarchical structure is predominant, few large community matrices that have been studied show *perfect* nesting or transitivity (Bascompte et al. 2003; Ulrich et al. 2014). Individual species pairs will in many cases deviate from strict transitivity, and in some cases these deviations should have effects on system stability. Current work has been intensely interested in such intransitivities, since under the assumption of symmetric species interactions, it is difficult to account for the prevalence of hierarchical structures and the coexistence of many species (Laird and Schamp 2006; Allesina and Levine 2011; Soliveres et al. 2015).

In our view, the explanation for these general patterns is nonadaptive selection on qualitative stability. Hierarchies accompanied by asymmetry of interaction strengths tend to approach qualitatively stability on their own and will be preferentially preserved regardless of how they may have come about.

The significance of qualitative stability for ecological networks may derive not only from the ease with which systems suitably structured find or attain highly stable states, but also from the freedom allowed to the action of other biological processes when the system is in or near a qualitatively stable state. In particular, individuals of species are regularly subject to selection on phenotypes that confer better performance, better defense, better competitive ability, and so forth, and such organismic-level evolutionary responses must be common, especially in large communities. Local evolution in a variety of individual fitness-related traits thus may alter the magnitudes of the interaction terms, even in relatively stable environments (Bolnick et al. 2011). In qualitatively stable communities, however, the magnitudes of these terms may be of little or no consequence for stability, so such communities are robustly stable in the face of considerable evolutionary change. Our suggestion is that nonadaptive selection results in recognizable network structures that typically lie not too far from these qualitatively stable "centers of concentration." The idea that observed community structures are feasible and stable subsets of randomly constructed networks is not new. However, previous literature did not address deviations from the symmetry assumption and thus missed a biologically meaningful class of hierarchical, asymmetric, one-side-dominated interactions.

A metaphor may be useful here. Imagine a kind of n^2-dimensional "stability landscape," in which peaks represent quantitatively stable, feasible, community configurations. Each point in the landscape consists of the values of the interaction coefficients in each position in the community matrix, and one dimension indicates stability. Some of these stability peaks will be isolated, far from their neighbors, but others will be in regions where the peaks are increasingly close together. In the limit, the peaks more or less coalesce in areas that form n^2-dimensional plateaus, sometimes of considerable size (areas of high QSS or qualitative stability). If you now randomly place communities in the positions of the peaks, those near other peaks are much more likely to find a new solution after a serious perturbation. Communities of isolated peaks will likely disappear at a greater rate. Therefore, over time we expect that the majority of existing communities will be found in areas near the plateaus and will be sparse elsewhere, regardless of how particular quantitative interactions among species may contribute to their stability at any one time.

7.4.5. A Conjecture

The two major, recurrent empirical patterns discussed in the ecological network literature—nestedness and modularity—are nicely interpreted as measures of qualitative stability. We have so far said little about modularity. *Modularity* is the division of the total community into weakly interacting, stable subsystems, and is widely understood to be a stable superstructure (Grilli, Rogers, and Allesina 2016). Nestedness, of course, corresponds to the roughly triangular hierarchical matrix that we have been discussing.

We suggest that modularity (independent subsystems) and nestedness (asymmetric hierarchical, pecking order–like modules) in some combination are the expected structure of communities after lengthy, ongoing nonadaptive selection on stability. Individual subsystems will tend to be hierarchical and asymmetric, but will exhibit weak or no interactions with other subsystems. As an idealized example, consider the two interaction modules in figure 7.6, one involving four species and the other three.

Figure 7.6A represents a fully transitive hierarchy, which could represent a competition hierarchy, with high levels of asymmetry. Species 1 outcompetes all others, species 2 outcompetes the remaining two, and so on (bold arrows). The light arrows indicate possible weak effects of the losers on the ultimate winners. (In competition, the ultimate loser may nevertheless get some of the resource that its competitor dominates.) This module will be in close proximity to qualitative stability if these thin lines are close to 0. Figure 7.6B represents a simple three-species trophic chain, which we

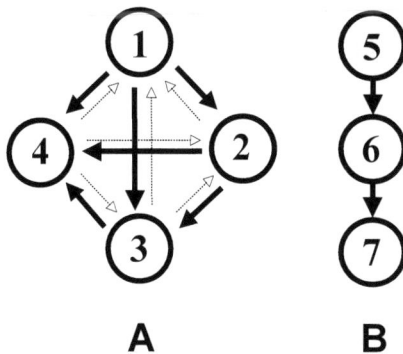

A   B

FIGURE 7.6. Two interaction modules discussed in text

$$\begin{bmatrix} - & - & - & - & 0 & 0 & 0 \\ \epsilon & - & - & - & 0 & 0 & 0 \\ \epsilon & \epsilon & - & - & 0 & 0 & 0 \\ \epsilon & \epsilon & \epsilon & - & 0 & 0 & 0 \\ 0 & 0 & 0 & 0 & - & + & 0 \\ 0 & 0 & 0 & 0 & \delta & - & + \\ 0 & 0 & 0 & 0 & 0 & \delta & - \end{bmatrix}$$

FIGURE 7.7. Matrix of pairwise interactions representing four-species and three-species modules depicted in figure 7.6

know from the previous sections is qualitatively stable. If the members of these two modules form a community but do not directly interact across modules, the combination will look like figure 7.7.

The two modules (shaded) will lie in a diagonal relationship, with 0s everywhere else. The ϵs represent small negative (or positive) values (the thin arrows in the previous figure); the ∂s represent small negative values. When the ϵ are all 0, this matrix as a whole is qualitatively stable, being composed of two qualitatively stable modules that do not interact. Of course, in real cases there may easily be small values of the ϵs, deviations from transitivity or from pure chains, and some interactions among the modules. All of these would tend to destabilize the community matrix. But communities structurally more similar to qualitatively stable ones will still more often be stable themselves.

This inference is the logical extension of our view that qualitative stability is a kind of "absorbing state." When a community ends up close to a qualitatively stable structure, it is comparatively difficult to change that structure, whatever the historical route the evolving community took to get there.

Our conjecture is, then, that nonadaptive selection acts upon ecological networks to increase the representation of either or both of the frequently observed *modular* and *nested* states. An article by Miguel Fortuna and colleagues (2010) asked the question, "Nestedness and modularity: two sides of the same coin?" Our answer is, "Yes, and the coin is nonadaptive selection on QSS stability."

7.4.6 Perspective

There is much to be learned about the conditions under which communities may exhibit feasible, stable equilibria (i.e., quantitative stability). We do not suggest that this traditional research is not important or productive. We are not saying that the asymmetric triangular structure of interaction is *the* general criterion for community stability and species coexistence. We cannot even evaluate to what extent the QSS values represent feasible equilibria or the way that communities evolve or assemble. We are making a different point: regardless of the range of stable community networks that could exist, the ones we are most likely to see are those that approach structures that are qualitatively stable; these will be the ones favored by nonadaptive selection. This fact will not change as our knowledge of quantitative stability conditions increases. In this case, as in others we have discussed, nonadaptive selection plays a significant role in explaining what we *see* even though it does not tell us what may be easy, difficult, possible, or impossible to construct.

In sum, high-QSS communities are difficult to lose or modify as a result of occasional episodes of evolution and perturbation. Isn't this actually what we want as a general definition of *stable*?

But that conclusion still does not address where the structural features leading to stability come from in the first place. Starting from a background of organismic natural selection, many researchers have emphasized the role of adaptive coevolution in producing just those structural features that yield community stability (Bascompte, Jordano, and Olesen 2006; Thompson 2006; Guimarães, Jordano, and Thompson 2011). However, often the exact coevolutionary mechanisms leading to the community patterns are not well characterized (see section 3.3). Is it coevolution alone that builds complex but nearly triangular structures that turn out also to be stable? Or is it also nonadaptive selection that lets us see only what is left, no matter how those structural characteristics originated? In fact, there is no reason that concerted, deterministic coevolution must bear the entire burden of forming the distribution of the stability-yielding patterns commonly observed among communities. Nonadaptive selection on community stability can change that distribution without any specific tendency of community networks to evolve on their own in a particular direction. Furthermore, if a somewhat stable network randomly moves toward a more stable state, nonadaptive selection can preserve that modification, just as any selective process can. Thus a long-lived community can experience a sequence of states over time, moving toward greater and greater stability, without a specific internal driving mechanism. Joan Roughgarden (1983) conjectured that coevolution itself was likely to cause extinctions, rather than build, bit by bit, stable structures, but also (consistent with our view) believed that the higher probability of extinction of less stable, more strongly interacting coevolving species might lead the community to more stable states.[10] So which is it? Do communities evolve from the "inside" or from the "outside?" The answer is likely to be: some of both (Hui and Richardson 2019). Nevertheless, the result in either case is still subject to further nonadaptive selection on stability. The fact that community structures actually seem to favor qualitative stability conditions suggests that nonadaptive selection has an explanatory role in the historical formation, as well as the maintenance, of ecological networks.

Finally, mathematical conditions for qualitative stability have not yet been fully explored. Qualitative stability is the strongest form of stability and requires no knowledge of the values of the interaction coefficients, except the signs. Between this extreme and quantitative stability there are known to be classes of stable matrices that are based on various degrees of partial knowledge of the values of the interaction coefficients (Logofet 1993, 2005). If some of these can be given plausible ecological interpretations, they may provide a foundation for additional stability structures upon which nonadaptive selection could work. They may be relatively rare, but we might be able to recognize them in nature as well.

7.5 Nonadaptive Selection and Predators Larger Than Their Prey

So far we have discussed community matrices, food chains, and food webs without specifying very much about general biological constraints that there may be on interactions among species. This has been the traditional approach, but as is well known, body size is related allometrically to a large number of physiological and ecological variables (Peters 1983; Calder 1984; Brown et al. 2004; chapter 6 in this volume). It has been clear for some time that realistic modeling of communities and their components should incorporate relevant information on size-related traits (Warren and Lawton 1987; Yodzis and Innes 1992; Berlow et al. 2004; Weitz and Levin 2006; Stouffer, Rezende, and Amaral 2011). When interacting species are of significantly different sizes, body-size allometries dictating differences in individual energetic needs as well as growth and reproductive rates have potentially significant effects on a range of community processes that can be related to stability. Currently, one of the most well-studied community-wide patterns involving allometry is the body-size ratio between predators and their prey (Pimm 1982; Woodward et al. 2005; Wooton and Emmerson 2005).

It is a commonplace observation that predators tend to be larger than their prey. The intuitive idea (e.g., Elton 1927, 59) that predators must usually be larger than their prey in order to capture and consume their victims, although true enough, is actually not fully convincing as an explanation of this pattern, because it hardly accounts for the generality and magnitude of observed predator/prey body-size ratios. We now know from large compilations that predators are usually *much* larger than their prey—often by orders of magnitude (Cohen et al. 1993; Brose, Jonsson, et al. 2006). This is true across major taxa and among both terrestrial and aquatic species.

Such a high predator/prey body mass ratio is understood to affect interactions of species pairs in food webs, in the following way. The per capita effect of a large predator on a population of very small prey may be reasonably large, but the converse per capita effect of the tiny prey on the predator's population will be very small (Pimm 1982, 132). Thus the difference in predator and prey sizes increases the *asymmetry* between positively and negatively signed interaction terms, which lie on opposite sides of the matrix diagonal. This asymmetry (and the stability it confers; see section 7.4.3) increases with the body mass ratio. With realistically high predator/prey body mass ratios, it is the effects on the carnivores of per capita changes in population numbers of the prey that would exhibit the smaller interaction values. In contrast, where predators and their prey are of similar size, we expect that this asymmetry would be greatly reduced or eliminated, contributing to instability. (Asymmetry would not necessarily disappear altogether, as other biological processes could come into play once the size-based asymmetry was effectively eliminated.)

Certainly increasing the interaction asymmetry may not be the only factor by which predator/prey body mass ratios affect stability and structure of food webs (Brose 2010). Numerous studies have modeled food web dynamics by incorporating directly allometric scaling effects of factors such as reproductive rates, growth rates, foraging efficiency, and metabolic requirements (Jonsson and Ebenman 1998; Emmerson and Raffaelli 2004; Loeuille and Loreau 2005; Brose, Williams, and Martinez 2006; Brose et al. 2019). Regardless of the specific modeling approach, high predator/prey body mass ratios emerge as potent stabilizing influences on both food chains and whole communities (Brose 2010).

Nonadaptive selection on system stability appears to have a major role to play in maintaining the prevalence of these relatively high predator/prey body mass ratios. As is the case for other nonadaptive selective processes, the globally constructed datasets sampling from a large sample of communities (Cohen et al. 1993; Brose, Jonsson, et al. 2006) reflect faithfully the ratios found in local communities. Observed predator/prey body mass ratios in any particular food chain may arise as a result of selection among consumer individuals for optimal rates of trophic energy transfers in food webs (DeLong 2020). As Charles Elton (1927) pointed out, there are upper and lower limits to the size of a consumer's prey. If too large, the prey cannot be handled, and if too small it may not repay seeking it out. So nonadaptive selection will tend to favor those ratios that are viable in a given interaction between predator and prey, and these will include many strongly asymmetric values. The stability conferred by such high predator/prey body mass ratios will guarantee that these are the ones we are most likely to see, as nonadaptive selection will tend to remove low values from our view, even if they are viable in the short run.

Quite apart from consideration of complex food web dynamics, a simple rule of thumb also highlights the role of nonadaptive selection in the maintenance of high predator/prey ratios. Imagine a pair of predator and prey populations whose dynamics are governed by exponential growth if grown alone, without an interaction. For stable coexistence of the interacting pair, the maximum per capita population growth rate (the Malthusian parameter r_{max}) of the predator must be lower than that of the prey. If not, then the predator population will outgrow the prey and eventually cause its extinction. The consumer would be growing faster than the resource and depleting it faster than it can be renewed. Thus a situation of equality or near equality of the maximum growth rates of predator and prey is parametrically unstable. Sooner or later small environmental or evolutionary changes would likely cause the system to remain in the region of prey extinction long enough for the prey to disappear. If in such a case the predator has a limited diet (or is even a specialist on the prey), extinction of both predator and prey is also a likely outcome. A faster-growing predator will ultimately cause complete extinction of slower-growing prey, even if it does not specialize on that prey species; it just might take longer.

Since r_{max} scales negatively with body size (Fenchel 1974), this suggests that predators must in most cases be larger than their prey, guaranteeing that they have a lower r_{max}. We do not often see predator/prey body mass ratios near unity or below, because such systems are eventually removed from our view by extinction of one or both participants when they do form. What we observe are mostly nonadaptively selected stable configurations: predators reproducing more slowly than, and interacting asymmetrically with, their prey.

We return to this topic, and a conspicuous exception, when we discuss macroevolution in chapter 8.

CHAPTER 8

Macroevolution: The Role of Nonadaptive Selection among Species

8.1 Introduction

We now turn to the role of nonadaptive selection at much larger scales of space and time. This level of evolutionary study is often referred to as *macroevolution*,[1] and it has overwhelmingly focused on the description of historical patterns. But there has always been a suspicion that we may be missing something with a purely historical account of macroevolution. Could there be causal processes at the macroscale of which we are currently unaware? Or are there ways to connect microevolution to macroevolution that provide explanations for historically observed patterns in terms of causal mechanisms, rather than just as historical sequences of events (cf. Gould 1998)? The older macroevolution literature proposed many now-deprecated evolutionary mechanisms, mostly as transformational alternatives to Darwinian natural selection for forming organismic adaptations. Such processes are of historical interest but are not the subject of this book. However, the more recent literature contains extensive discussion of a *selective* process involving whole species and its potential explanatory role in macroevolution, under the broad banner of *species selection*. Some versions of this topic envision a theory that would be able to explain certain macroevolutionary patterns solely on the basis of demonstrable, mechanistic, and necessary relationships between species characters and species-level fitness. This is directly relevant to the theme of this book, and we propose that the concept of nonadaptive selection provides a novel perspective that may clarify the issues and provide a productive way forward in this research area.

8.2 Species Selection

The essence of species selection is that clades of species are "populations" within which there is a selection process among species, based on systematic variation in speciation (= "birth") and extinction (= "death") probabilities of whole species—that is, a species-level "fitness." The distribution and relative abundances in the clade of different "kinds" of species change as a result of this selection. Different views of species-level characters are ultimately related (with varying significance in different versions of species selection) to the traits of the lower-level entities within them, such as organisms.

Many researchers have considered species selection (or "sorting") to be mostly a description of evolutionary history. Others have taken the loose analogy with Darwinian natural selection more literally and claimed that species selection was only valid or interesting when higher-level *adaptive* evolution was occurring at the species and clade level. Currently, species selection per se is for many a part of an eclectic, "sorting" framework that includes all historical sorting processes, including (mostly?) historically descriptive species selection, widespread directional environmental change, episodes of mass extinction, and effects of interactions with distantly related clades (Rabosky and McCune 2010; Jablonski 2017). However, this diverse conceptual structure is not exactly the causal selection theory for which many researchers had originally expressed hope. See appendix 10.13 for a fuller discussion and documentation of the historical development, conceptual issues, and current status of species selection.

8.2.1 Selection, but No Adaptation at the Species Level

Selection, as defined by the Price equation, is certainly consistent with species selection. As long as the "population" of species can be defined and circumscribed, and it is possible to assign relative "fitnesses" to the different species' "phenotypes," selection will occur from one time period to the next. So there is certainly a potential *selection* process among species of a clade; however, here the resemblance to adaptive natural selection among organisms largely ends. Simply by being an example of some kind of selection, it does not follow that there ever is *adaptive* selection at the species level, nor does the observed selection process necessarily provide any general explanation for repeated patterns of clade change.

It is relatively easy to see why *selection* among species is straightforward, but adaptive evolution at the species level encounters insuperable obstacles. In order for *adaptive* evolution to occur, units under selection have to share a common, local environment (see chapter 3). Clade membership is based on genealogy, not the occupancy of such a shared environment. The species of a clade cannot undergo adaptive evolution at their level, because there is nothing in particular for them to adapt *to* (Damuth 1985; Grantham 1995). Nonadaptive selection, in contrast, does not require a shared environment, because adaptation is not happening.

Selection without adaptation, of course, is the topic of this book, so it is tempting to say simply that *species selection is nonadaptive selection* and leave it at that. However, this will not yet lead us to a causal theory, because of another implication of the absence of a shared environment.

To have a chance of providing mechanistic, rather than ordinary historical, explanations of clade patterns, we must have some reason for regarding attributes of the species (at any level) as *reliable* for inferring species-level fitness. That is, as usual, the characters have to be associated with "heritable" variation in fitness (Jablonski 1987). In organismic selection, the shared environment guarantees that a given phenotype contributes the same expected effect on fitness, regardless of the individual bearing it (see the discussion in chapter 3). Where there is no shared environment, this connection is broken; populations within a species, and species within a clade, do not necessarily experience the same conditions, so they are not necessarily associated with the same fitness effects (see appendix 10.13).

However, we can use the fact that species and clades are historical constructs to suggest a role for nonadaptive selection processes, at ecological levels, in resolving the problem of the absence of a shared environment.

8.2.2 *Historical Constructs and Nonadaptive Selection Processes*

Species and clades are global constructs, composed of samples of the ecological units that make them up. As we have seen in chapter 2 and elsewhere throughout this book, one way that we can be sure that a global construct represents the same macroscale pattern that its components do is when that pattern is the result of a nonadaptive selection process.

In other words, nonadaptive selection results in a lawlike relationship, whereby subcomponents of a global construct are all effectively the same in a particular respect, regardless of variation in space and time (recall the triplet distribution in communities or the energy equivalence relationship within trophic levels in chapter 2).

So, by considering species selection to be a process whereby *nonadaptive* processes underlie differences in species-level fitnesses, we solve the major outstanding difficulties we have identified in previous sections and discuss more fully in appendix 10.13. Selection can involve any characteristics attributable to the species. Those characteristics can be associated with characters at the level of ecological units within the species, whose implications are unaffected by ecological variation throughout the species. That is because they operate by means of a lawlike generalization that is nonadaptive. The lawlike mechanism *does* cement the characteristics of the ecological units involved in it with species-level fitness differences. This obviates the need for there to be a "shared environment" for species selection. And the nonadaptive, lawlike mechanism describes not only the mechanism that does this but also the domain, or circumstances in which we might expect it to occur.

We describe a concrete example of this nonadaptive approach to species selection in the next section.

8.3 Elevated Extinction Rates for Large Hypercarnivores

In section 7.5 we noted that ordinarily, predators are *much* larger than their prey (Brose, Jonsson, et al. 2006). We pointed out that this regularity derives—at least in part—from a nonadaptive selection mechanism based on the relative stability of predator-prey systems that differ in their predator/prey body size ratios (see Brose, Williams, and Martinez 2006).

However, there is one conspicuous exception to this general pattern. At the large end of terrestrial body sizes, mammalian predators tend to be of similar body size to, and are sometimes even smaller than, their prey. Moreover, this configuration appears to evolve repeatedly throughout the fossil record of mammalian carnivores, with members of successive clades evolving to fill these large carnivore/large prey niches, while the previously dominant hypercarnivore clade (or clades) declines (Van Valkenburgh 1991).

Terrestrial mammalian predators differ in the degree of meat in their diets, and those whose diets are almost entirely composed of vertebrate flesh are called *hypercarnivores*. Both a hypercarnivorous diet and the body mass of extinct species can be inferred for fossil species from the carnivore skull and dentition (Van Valkenburgh 1991; Van Valkenburgh, Wang, and Damuth 2004).

We do not usually have direct knowledge of the prey of extinct predators, but in the case of *large* hypercarnivores we can make a strong inference about their prey's relative size. From energetic considerations, carnivores above about 21 kg in body mass must transition from consuming primarily invertebrates to specializing on large vertebrate prey; foraging costs and required rates of intake make it impossible for a large hypercarnivore to make a living from resources that come in small, dispersed packages such as insects and small vertebrates (Carbone et al. 1999; Carbone, Teacher, and Rowcliff 2007).

So from this we know that any terrestrial mammal above approximately 21 kg in mass and actively consuming almost exclusively an animal diet *must* be a hypercarnivore *and* restricted to relatively large-size prey. This constraint is based on principles that apply to any mammalian (or mammal-like) carnivore, at any time, regardless of the way that it lives or the particular species that make up its diet. (Dinosaur top predators were presumably also subject to at least some of the same kinds of foraging and metabolic constraints. However, our current understanding of predatory dinosaur physiology and ecology is still highly uncertain [Marshall et al. 2021; Farlow et al. 2023]. At this writing, we cannot know for sure whether predatory dinosaurs were in the domain of the large-hypercarnivore effect described here for mammals.)

As we saw in chapter 7, moving from a typical predator-prey food web component—in which predators are much larger than their prey—to one in which the predators are of equal or somewhat smaller size than their prey, substantially *lowers* the asymmetry of the species interactions. Asymmetric interactions lead to stability, because the matrices describing them approximate more closely structures that confer qualitative stability (see section 7.4). A relatively low-stability system with *symmetric* interactions would need to be more delicately tuned to avoid straying into unstable territory. Populations of large hypercarnivore species should thus tend to exhibit shorter durations and higher extinction rates than those of other carnivores, and the same should be true of clades that have many species consisting of large hypercarnivores.

The detailed fossil record of extinct North American canids (the dog family) documents just this pattern (Van Valkenburgh 1991; Van Valken-

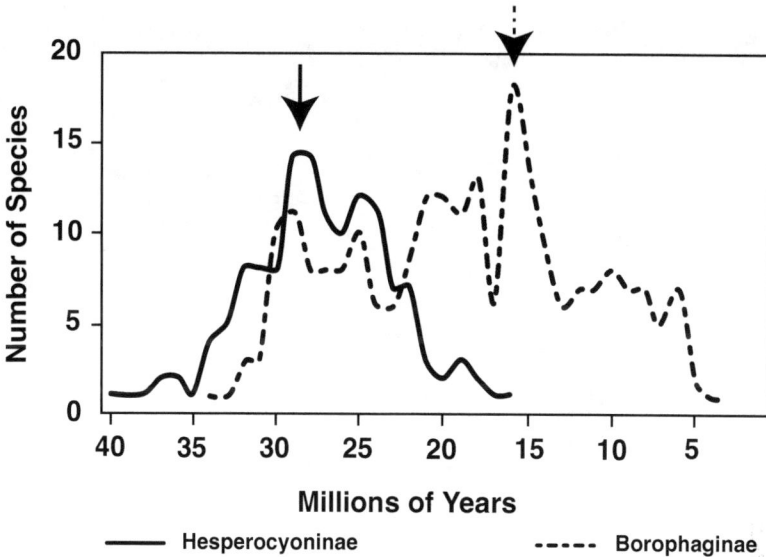

FIGURE 8.1. Species richness of extinct subfamilies of North American canids, spanning the last 40 million years. Arrows represent time at which large hypercarnivores arose in each clade. Figure based on Van Valkenburgh, Wang, and Damuth (2004).

burgh, Wang, and Damuth 2004). The two extinct canid clades, the sub-families Hesperocyoninae and Borophaginae, rose and fell in a remark-able pattern (see figure 8.1). Early members of both clades are relatively small, generalized predators, but throughout each clade's history there is a gradual increase in body sizes of most of its lineages and in the numbers of species. However, at the moment that large hypercarnivores appeared in each clade and began proliferating, the clade began a distinct decline. Data show that large hypercarnivores exhibited statistically significant shorter species durations than nonhypercarnivores (Van Valkenburgh, Wang, and Damuth 2004). The time from the high point of species rich-ness to the effective extinction of the clade—the period of time during which lineages of large hypercarnivores were present—was in each case approximately 10–15 million years.

Few other carnivore groups have as complete and well-studied records as these canids, but it appears that this pattern is widely observed. The other terrestrial top carnivore clades, such as the cats and catlike forms (es-pecially including sabertoothed forms), also exhibit this iterative evolution

MA

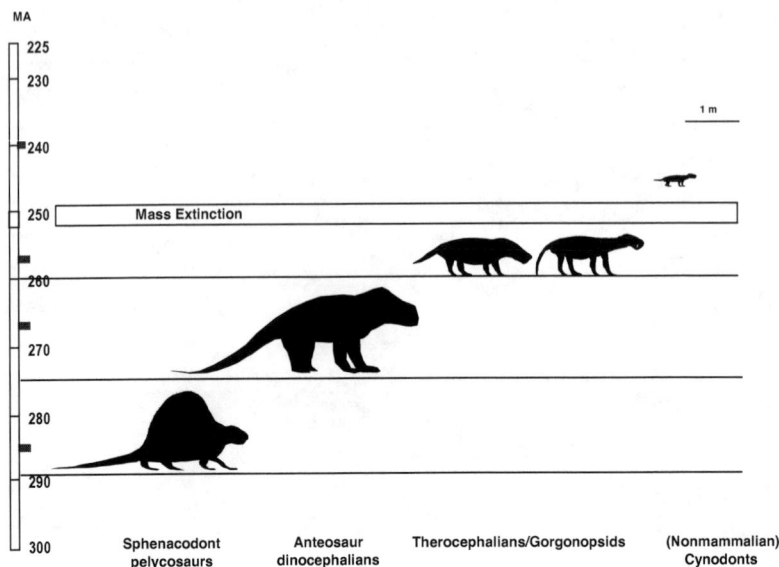

FIGURE 8.2. Nonmammalian synapsid top carnivores throughout the Permian. Successive clades containing probable hypercarnivores dominate for periods of approximately 10–15 million years, similar to the extinct canids of the Cenozoic (figure 8.1). The extremely severe Permo-Triassic mass extinction cut short the dominance of synapsid large predators. Triassic forms never really crossed the threshold into hypercarnivory, and the role of dominant terrestrial predators was ceded to theropod dinosaurs. MA = Millions of years before present. Figure based on Van Valkenburgh and Jenkins (2002).

of hypercarnivory and high extinction rates for large hypercarnivores (Van Valkenburgh 1999, 2007; Piras et al. 2018; Balisi and Van Valkenburgh 2020). In fact, there is a hint that the same mechanism may have been operating almost 300 million years ago in the Permian period. The first large terrestrial carnivores arose in the Permian, among the synapsids (the major vertebrate clade within which lies the ancestry of mammals). Three successive dynasties of these nonmammalian synapsid large hypercarnivores filled the top predator niche, each maintaining its dominance for a duration of approximately 10–15 million years, similar to what we see in the Cenozoic canids (see figure 8.2) (Van Valkenburgh and Jenkins 2002).

We believe that a nonadaptive mechanism based on the instability of symmetric species interactions is likely sufficient to generate this widely observed bias against the survival rates of large hypercarnivores, without requiring any unfamiliar evolutionary processes or special historical circumstances. Size increase may be favored by adaptive selection within lineages, for many reasons including access to new resources. Specialization in diet can also occur via conventional organismic selection involving multiple causes (Futuyma and Moreno 1988). At any given time individual adaptedness among the hypercarnivores may be quite high, and there is likely no general selective value for individuals to revert to previous ecological niches, especially if they have lost morphologies advantageous for a niche previously occupied by their species. The effect of the evolution of large carnivores on clade decline appears to take a long time, suggesting that overall it is weak, but it is always there.

The fact that this very general pattern is associated only with the *combination* of large size and hypercarnivory suggests a specific nonadaptive mechanism within the domain of terrestrial carnivores, which involves those two characteristics. Neither large body size alone, nor specialization by itself, is consistently related to short species durations, among carnivores or animals in general (Tomiya 2013; Day, Hua, and Bromham 2016; Balisi and Van Valkenburgh 2020). Likewise, although special historical circumstances, such as widespread ecological changes or competition with distantly related clades, may affect species durations, they evidently do not overwrite the effects of the nonadaptive process we have identified.

So with respect to large terrestrial hypercarnivores, we have a nonadaptive selection process that operates on every population of a species, resulting in an increase in long-term instability. Thus, the characteristic of the whole species (a historical construct) is *reliably* similar and inherits the instability that is characteristic of populations of the species. In other words, here we have a macroevolutionary pattern, caused by a demonstrable ecological mechanism, that links micro- and macroevolution but is *not* an example of Darwinian adaptive microevolution somehow extended in time. Neither is it caused by a purely macroevolutionary process that springs up de novo "at the species level" that has no connection to the ecology of the species' populations. If the causal process is nonadaptive selection, it is as valid to say that the species-level (or clade-level) pattern is caused by the nonadaptive selection process as it would be to say that the lower ecological (population or community) patterns were caused by the same nonadaptive selection process.

This seems, in fact, to be what many people have implied that they wanted out of species selection. It may not be what most people expected, but we think it is a causal mechanism that works.

8.4 Outlook for Selection-Based Macroevolutionary Explanations and Laws

The change in perspective from an adaptive to a nonadaptive view of species selection comes with both costs and benefits. First, this move implies that there are no novel adaptive processes at macroevolutionary scales that act to optimize or increase the "fit" of species or taxa (as units) to their circumstances; there is no species-level adaptive safety device that will further improve (for organisms) the results of adaptive evolution among organisms. This is already a widely held view, but the language in which macroevolution is often discussed is nevertheless still heavily laden with adaptive references and imagery (e.g., "adaptive radiation," "adaptive zone"). Should anyone develop a coherent, convincing, and fruitful concept of adaptation for taxa as units, it may not look very much like today's species selection, but it would presumably complement the *nonadaptive* explanatory schemes that we think are currently a more likely way to make progress.

Second, we suggested earlier that more explicit nonadaptive mechanisms for explaining species selection patterns will not lead to *simple* generalizations about macroevolution *as a whole* (e.g., "Specialization is a dead end"). Such generalizations tend to not be very successful (Day, Hua, and Bromham 2016). Rather, successful macroevolutionary selective explanatory schemes will likely resemble more detailed and specific generalizations (e.g., "For top carnivores in terrestrial ecosystems, forced specialization to feeding on prey of similar body size to the predator leads to elevated predator extinction rates and clade shrinkage"). We see this as a hopeful sign; what we lose in breadth of applicability, we gain in precision and confidence in the lawlike explanations for more specific phenomena. We anticipate that seeking the explanatory power of nonadaptive mechanisms in macroevolution may allow the field to expand through proliferation of such explanations for specific patterns. Regarding species selection as a nonadaptive selective process identifies what it is that we should be looking for when we want a causal explanation of macroevolutionary patterns. We are looking for demonstrable lawlike processes, likely involving nonadaptive selection, which fortunately, as we argue, may be more common than people have realized.

Recognizing that taxa are historical, genealogical constructs significantly expands the potential scope of explanations of macroevolutionary patterns. Now nonadaptive selection processes at any level, and particularly ones involving ecological entities, may be visible as reliably giving rise to fitness-related, species-level properties. At the same time, members of taxa may be subject to many contingent influences that may impact speciation and extinction rates, but entire taxa are likely subject to relatively few lawlike nonadaptive selection processes that do so.

Macroevolutionary "laws" or generalizations that are directly tied causally to ecological "laws" operating at lower, observable levels form the basis for a workable species selection. All of these "laws" are based on nonadaptive selection processes, and within their domain(s) they have the claim to universality. Yet they coexist peacefully with Darwinian adaptive evolution at the organismic level and with adaptive evolution at any other levels at which adaptive evolution may occur. This is possible if we give up the unrealistic expectation of the dominance of local adaptation at all levels.

As we have shown in previous chapters, loosening adaptation's grip on our views of the evolution of biological systems above the organismic level allows explanation of a wide range of different kinds of recurrent patterns by means of nonadaptive selection. Loosening the grip all the way up the hierarchies of life should similarly allow us to explain macroevolutionary patterns caused by nonadaptive selection among species, regardless of the level(s) at which the relevant nonadaptive selection processes are acting.

We end in full agreement with Rabosky (2013, 497): "The past few decades have demonstrated that the causes of macroevolutionary rate variation are complex. We are still grappling with the meaning of speciation and extinction as measured over macroevolutionary timescales. We are beginning to move beyond the conceptual limitations of highly phenomenological birth-death models to recognize that macroevolutionary rates are not causal agents of large-scale diversity patterns, but rather summary statistics of the collective behavior of interactions at lower levels of biological hierarchies." The way to reconcile general macroevolutionary patterns with interactions at lower levels—and maintain mechanistic explanations—is through nonadaptive selection.

CHAPTER 9

Summary and Perspective

In this book we have gone into some detail to explain a relatively small number of instances in which we believe the effects of nonadaptive selective processes are evident. It is clear that this concept applied to biology is not a "theory of everything." But the question then arises, is the exercise worth the effort? In this book we are not arguing for a general reorientation of existing research traditions. Rather, we hope we have shown how recognizing the properties of a particular class of nonadaptive processes opens up additional ways to make progress on specific questions of explanation in ecology and evolution, particularly at large ecological scales. But what do we really get for adding nonadaptive selection to our explanatory toolbox in ecology and evolution? In a word, we get *laws*.

We have shown that nonadaptive selective processes can form the basis of lawlike statements that have the properties that are widely ascribed to "laws" in current philosophy. We do not argue that nonadaptive selection is necessarily the only source of such biological laws, but nonadaptive selection may be one area in which they are most easily discerned. One should not underestimate the significance of recognizing genuine laws, or lawlike statements, in ecology and evolution and in biology in general. Because of a formal, but now questionable, view of natural laws as existing only in the physical sciences, biology has long been considered to have no laws. Rather, biology is widely considered to be concerned merely with particulars of the biological systems that have evolved over the earth's long history. To be sure, biology has theory and processes specific to those systems, but it usually makes no claims that approach universality.

Might it be true to say that all laws or lawlike generalizations are non-adaptive, whether selective or not? Only something that is nonadaptive is divorced from the local environmental context and can be a firm basis for generalization. If so, the part of biology that is most like the other sciences is nonadaptive. It is often said that theoretical biologists suffer from "physics envy" (Cohen 1971), but that should not be taken to mean that we should ignore or devalue those biological phenomena whose most effective explanations resemble more closely the lawlike statements of physics than they do the special results of adaptive processes. If we have lawlike explanations for repeated, widespread phenomena, then we should be happy to use them.

9.1 Life Elsewhere in the Universe

Domain-specific universal laws may even be of limited practical value in future investigation of life elsewhere in the universe. We expect that soon astronomers will discover exoplanets that plausibly support life (Cockell 2020; Arthur 2023). For many reasons extrasolar life is most likely to be simple, microbial-like, rather than complex, multicellular "animals and plants" (Ward and Brownlee 2000). However, if in fact we do discover complex life, we potentially already know many things about it, depending on the law domains represented. So not only may we be certain of some structural aspects of exocommunities and species interactions, physiological and ecological allometries, but a consideration of known laws could guide us toward the more important questions to ask (Colyvan, Damuth, and Ginzburg, 2019).

9.2 The Origin of Adaptation and the Zero-Force Law

Daniel McShea and Robert Brandon (2010) identified and discussed a general biological law, the Zero-Force Evolutionary Law (ZFEL) (see also Brandon and McShea 2020). The ZFEL states that living systems that exhibit variation and heredity (of the kind we see in adaptive evolution) cannot help but produce and accumulate variation, whether they are subject to selective forces or not. This underlies the mechanism of Darwinian adaptive evolution that characterizes most organismic life on our planet.

So where did adaptive evolution and the ZFEL come from? Nonadaptive selection characterizes nonliving systems, as we see in appendix 10.8. Therefore, the origin of adaptation must have occurred by means of nonadaptive selection. It is easy to see why.

The major requirement for the building of adaptations is the characteristic of life that consists of reasonably (but not entirely) faithful replication in reproduction, with the ability to conserve valuable mistakes long enough for such variation to contribute to adaptive evolution. This leads to local adaptation and the diversity of life and is the requirement for the ZFEL to be in force.

Nonadaptive selection enters the picture as follows. Organisms with reproduction/replication systems that are too ineffective tend to produce many failures and have low heritability of even advantageous mutations. Organisms that for some reason have perfect replication, on the other hand, would never produce any variation and thus cannot adapt to anything. So we have not seen, and will never see, either of these extremes, because even if they appear they either fail to function or fail to adapt and diversify. The origin of adaptive evolution itself can thus be seen as the result of nonadaptive selection on such systems, and so can the ZFEL that undergirds adaptation. One of the consequences of this is that it should be comparatively easy for the process of adaptation to originate anywhere in the universe, as it depends on the outcome of a very general set of nonadaptive requirements that does not presuppose any of the specific machinery by which variation and heredity is supported in our own biosphere (cf. McShea and Brandon 2010, 133–34).

9.3 Populating of the Biosphere with Stable Structures

Our argument suggests that the products of nonadaptive selection are all around us in biological systems but are seldom identified as such by biologists. Richard Dawkins (1976) provided a beautiful description of nonadaptive selection among "proto-life" but then switched to considering only adaptation after life "begins."[1] Partly, the explanation for this oversight may be simply that stable configurations are more long lasting and thus available for us to observe, while unstable ones are ephemeral and mostly gone by the time of our observation. The products of these processes tend to be regarded as the stable parts of biological reality. Variation and diversity appear to be the things that need historical and adaptive explanation. But the stability goes unnoticed or is taken as given. The absence of what is *not* seen gets little attention. Yet a consideration of both observed actualities and theoretical possibilities can sometimes reveal the explanation for the stable scaffolding itself. We can agree with Arthur Eddington (1928, 266–67) that "the contemplation in natural science of a wider domain than the actual leads to a far better understanding of the actual."[2]

As pointed out in our example of food web motifs (see section 2.2), only four relatively stable triplets have been generally noticed in ecology and have been given generally recognized names; many aspects of their dynamics are understood (Holt 1977; Polis, Myers, and Holt 1989). Likewise, it has long been thought—and some evidence suggests—that most speciation events (in the sense of reproductively isolated subpopulations) are ephemeral and largely go undetected (Mayr 1963; Allmon 1992; Rosenblum et al. 2012; Rabosky and Matute 2013). Only if the new species is stable enough to survive do we have any reason to name it, or the ability to count it in biodiversity studies, or the possibility of observing it in the fossil record. Clearly there is a potential selective process occurring, but we know little about it as such. This is not different from astronomy, in which we give names to and study stars, planets, and their moons. Our attention is attracted to structures that "exist," that are stable, and is directed away from ephemeral unstable or less stable configurations, so selective processes involving them go unnoticed. Yet these stable structures of nature are the scaffolding within which unique and contingent events and processes take place. Simply put, nonadaptive selection processes imply that at some specific scales and circumstances, we *can* "replay life's tape" and get many results that will be qualitatively the same. This does not discount the importance of contingency but suggests that the stable structures within which contingent processes operate are more highly constrained (Sepkoski 2016).

9.4 Conclusion

In this book we have demonstrated the difficulty of explaining large-scale patterns in ecology and evolution from a perspective committed exclusively to an adaptive selective framework. By linking the structure of food webs, communities, and the maintenance of ecological allometries with traditional macroevolutionary topics, we embrace what have usually been thought of as separate fields of inquiry, but all of which have had an uneasy relationship with mainstream evolutionary theory that concentrates on local adaptation. To the extent that these fields, approached from the perspective of nonadaptive selection, can be mutually supporting, we can hope that advances in one will contribute to advances in the others. Moreover, we expect that the nonadaptive perspective will also be applicable to some degree to processes operating below the level of the individual. For example, this perspective may apply to fields such as systems biology, where complex systems and network thinking have come to prominence and the emphasis is on explaining general properties of biological systems rather than their adaptation to particular environments (Proulx, Promislow, and Phillips 2005; Boogerd et al. 2007). As always, the value of applying a general concept or theory across scientific disciplines comes not from recognizing the logic of the extension but from finding that it is useful to have made it.

Finally, there is more than just practical value to the delineation of biological laws. In a celebrated passage about the new uniformitarian geology of his time, Charles Lyell (1830, 166) wrote: "Thus, although we are mere sojourners on the surface of the planet, chained to a mere point in space, enduring but for a moment of time, the human mind is not only enabled to number worlds beyond the unassisted ken of mortal eye, but to trace the events of indefinite ages before the creation of our race, and is not even withheld from penetrating into the dark secrets of the ocean, or the interior of the solid globe; free, like the spirit which the poet described as animating the universe." Lyell here implied a stronger claim than just the ability to generalize about observations restricted to our own time, or our own planet. We can do more than explore what exists or has existed in one specific historical chain of events. We argue that, indeed, it is possible to regard ecological and evolutionary "laws" as expressing a kind of universality, valid wherever in space and time their domain applies. So not only do we wonder at the richness and diversity of earth's particular natural history, but armed with what is so far only a meager set of ecoevolutionary laws, we can already catch a glimpse of infinity.

Appendices

10.1 Introduction to Allometry (Power Laws)

10.1.1 The Basics

Throughout this book we make frequent references to allometric relationships (known more generally as power laws). Power laws express a relationship between two variables in which one of them varies as a power of the other. They have the general form

$$Y = aM^b,$$

where Y and M are variables of interest, and a and the exponent b are constants. Power-law relationships are widespread in natural and human-built systems and can be generated by a variety of common types of processes (Caldarelli 2007, 84ff.). Of particular relevance to organismal biology are situations in which one variable varies as a regular multiple or percentage of another.

In biology, variables that change (or *scale*) with organism size very often follow a simple power-law relationship, with M (as above) being a variable representing size (frequently, body mass). For historical reasons, we often refer to such power laws as *scaling relationships* or *allometries* (Gould 1966). Originally, *allometry* referred to patterns of relative growth of different morphological traits in organismal development (Huxley and

Teissier 1936), but the term is now widely applied to power-law relationships in other areas of biology.

For convenience and ease of visualization, we often transform the allometric or power-law equation to logarithms:

$$\log Y = \log a + b(\log M).$$

This equation means exactly the same thing as the first, but the transformation makes it take on the form of an equation of a straight line. If one plots $\log Y$ against $\log M$, then points conforming to the equation form a straight line, with $\log a$ as the Y-intercept and b as the *slope*. The advantage of making this transformation is that it simplifies finding and visualizing b, using a linear fit to empirical data points in order to estimate the value of b. Furthermore, the scatter in such a logarithmic plot now represents *proportional* variation, which is more relevant to most areas of biology, in which multiplicative processes dominate (Galton 1879; Kerkhoff and Enquist 2009).

Thus, in the text we often call some variable an exponent in an allometric relationship, but illustrate it with logarithmically transformed data where the exponent is the slope of a line. We tend to treat slopes and exponents as the same things, always assuming that the slopes refer to the exponents but as represented on a logarithmically transformed plot. Often we will not be very interested in the intercept ($\log a$), since it is the slope, b, that tells us how things are proportionally changing across size.

An important special case deserves mention. When the exponent of M is zero, the value of Y is a constant for all body sizes. In this case across all values of M there is no systematic change with body size. The variable is independent of body size, and we also sometimes refer to this allometric relationship as a body size "invariant."

Note that sometimes we are interested in size-scaling variables whose values are derived from a combination of more than one allometric relationship. For example, perhaps we have a variable Z that is the product of Y and X, both of which scale with body mass. We can obtain the allometric exponent for Z by summing the exponents of X and Y:

$$Z = YX = a_y\, M^{b_y}\, a_x\, M^{b_x} \propto M^{b_y + b_x}.$$

Likewise, dividing one relationship by another results in subtracting exponents. Such derived exponents will be strictly correct only if deviations of the values of Y and X from their respective allometric relationships are independent of each other. Independence is often approximated in many datasets but is not guaranteed. In situations in which the exponents of component variables sum to zero, one may say that the original exponents "cancel" each other and form an invariance, as in the case of energy equivalence described in section 2.2 and that of the scaling of lifetime reproduction (see chapter 5).

10.1.2 Ecological Allometry

Some allometric relationships (particularly in ecology) are associated with a considerable amount of variation around the line defined by the log-transformed allometric relationship. In most cases this does not represent observation error, but real biological variation (statisticians call both kinds of deviations "error"). For example, animal population densities vary over several orders of magnitude at any one body size, because habitats exhibit different levels of primary productivity, different distributions of resources, and so forth. Furthermore, body size affects all aspects of a species' physiology and ecology; it is impossible to find a large dataset spanning a significant size range in which the species differ only in size and the one variable of interest. Additional observations will not remove variation due to such causes. Such high levels of variation cause two problems.

First, they complicate the empirical fitting of a line to the data. In fact, there is no general way to obtain a "best fit" line for data without making specific assumptions about the relative magnitudes of the error variances of the two variables (e.g., Y and M in the allometric equation; Kendall and Stuart 1979). Almost always these error variances (which refer to all of the deviations from the true line, observational and biological) cannot be directly measured (since that would require that we already know the true line of best fit!). However, one can make reasonable inferences about their relative magnitudes and use a curve-fitting technique that is appropriate (Kuhry and Marcus 1977; Damuth 1993; Legendre and Legendre 1998). When error variation is high, different assumptions about the error variances and heterogeneity among the species in correlated characters result in uncertainty about the true value of the exponent b. In most cases we must accept that what we derive empirically is only an approximation of the true value, and that we will never settle disputes about small differences in exponents by recourse to an analysis of a dataset.

Second, a high range of interspecific variation in a trait at a given body size means that if one looks at a relatively small range of body sizes, one will not necessarily find the same (if any) scaling relationship as one finds in the full dataset spanning many orders of magnitude in body mass. The reason is obvious from simple inspection of the geometry of the bivariate plots. If one removes most of the variation in body size, it is unlikely that one will recover a consistent pattern over the size range that is left. Too often, researchers assume that if a relationship is observed at a large scale, it will continue to be evident at any arbitrary lower scale, as if the ecological allometry were somehow fractal (e.g., Bohlin et al. 1994; Dahlsjö et al. 2015). But failure to find the relationship at a very small scale is no evidence that the overall pattern is suspect or nonexistent. There is just too much noise at the smaller scale to pick up the larger-scale pattern.

10.2 Qualitative Stability Conditions

The necessary and sufficient conditions for an $n \times n$ matrix A with elements a_{ij} to be qualitatively stable (Quirk and Ruppert 1965; ecological interpretation in May 1973b) are:

1. $a_{ii} \leq 0$ for all i.
2. $a_{ii} \neq 0$ for at least one i.
3. The products $a_{ij}a_{ji} \leq 0$ for all $i \neq j$.
4. For any loop of interactions of three or more indices, the product $a_{ij}\, a_{jk}\, a_{kl} \cdots a_{mi} = 0$.
5. The determinant of the matrix, $\det A \neq 0$.

The first four conditions allow simple biological interpretation, but we cannot offer an intuitive interpretation of the biological significance of condition five, which is both quantitative and technical. The first two conditions describe intraspecies effects. The first says that there is no Allee-effect-like acceleration of population growth caused by the species' abundance. The first and second conditions jointly specify negative density dependence for at least one species. The third excludes symmetric competition and mutualism, $(-, -)$ and $(+, +)$ respectively, but does not exclude strong asymmetric interactions, $(-, 0)$ and $(+, 0)$. The fourth forbids non-zero loops of interactions of three or more species. Interactions have to be limited to just pairs. May (1973b) provided good examples of qualitatively stable and unstable ecological structures. It seems at first sight that mutualism and competition are not conducive to stability, but May did comment in this article that these interactions are commonly commensalism $(+, 0)$ and amensalism $(-, 0)$. We absolutely agree and thus stress the example of the triangular matrix, which is a common ecological case. May also mentions Mark Williamson's (1972) argument that predation is often closer to commensalism, $(+, 0)$, and we have also found this to be quite plausible (Arditi and Ginzburg 2012).

Condition five raises a technical issue that is beyond the scope of this book. Dmitri Logofet (1993) suggested a qualitative analog of this quantitative condition that is also quite technical. Thus, when interpreting the structures we observe in biological networks, we can rely on just relations one through four as the criteria for qualitatively stable outcomes; condition five will not concern us. In other words, when evaluating actual interaction matrices observed in nature, we seek evidence that one or more of the first four conditions are met and, when they are, we infer that they are contributing to the stability of the system represented by the matrix and therefore may be a target of nonadaptive selection.

Logofet's (1993) book contains an interesting collection of stability concepts that are intermediate between complete qualitative stability and general quantitative stability (see section 7.4.6).

A substantial generalization of qualitative stability is quasi-sign stability (QSS), introduced by Stephano Allesina and Mercedes Pascual (2008). Computer-generated random numbers allow an objective determination of intermediate levels of stability for interaction matrices when the signs of interaction coefficients are known. We first discussed this concept in section 2.2, and appendix 10.12 contains detailed information and many examples. If nonadaptive selection provides constant pressure against unstable configurations, we expect to see in nature many structures with relatively high QSS values. This seems to be the case as shown in various examples in this book.

10.3 Allometry and Energy Equivalence (Damuth's Law)

10.3.1 The Pitfalls of Using the Population Density Proxy and How to Avoid Them

As discussed in chapters 2 and 6, energy equivalence is a fundamental invariance that is likely maintained by nonadaptive selection within trophic levels of communities. Population energy use can be broken down into its components, but if we are right, what is causal is simply the rates at which populations use energy. Unfortunately, direct measures of population energy use are rare. It is much easier to find published population densities than population energy-use values, and as described in the text, to compare the scaling of density with the likely scaling of individual metabolic rates (see chapter 2). When energy equivalence holds, all other things being equal, density should scale as does the exponent of individual metabolism, but of opposite sign. As long as the distribution of energy (or resource density) is independent of body size, ordinary processes of population dynamics do not significantly overprint the relationships among energy use, density and size (cf. Ginzburg and Colyvan 2004; DeLong and Vasseur 2012; Yeakel, Kempes, and Redner 2018; Hatton et al. 2019). Density scaling is thus serving as a rough proxy for population energy use. Multiplying population density by corresponding basal or resting metabolic rates results in the expected independence of energy use and body mass in large and diverse global samples (Charnov, Haskell, and Ernest 2001; Ernest et al. 2003). However, these energy-use values would represent resting metabolism under laboratory conditions. What little direct data are available support the idea that field metabolic rates, though variable, roughly parallel basal/resting metabolic rates (Nagy, Girard, and Brown 1999; Nagy 2005), and that energy budgets for individual species populations show no relationship to body mass (Humphreys 1981; Damuth 1987; Hechinger et al. 2011). So our assumption that actual energy use approximates a multiple of basal metabolism is relatively solid.

We are using the *scaling* of population density to evaluate the scaling of population energy use. This is not the same as studying population density itself; we are not interested here in using the allometry of population density to explain or predict population abundances per se. This perhaps surprising claim derives from the original discovery that among herbivorous mammals population densities scale such that population energy use is independent of body mass (Damuth 1981b). Thus energy equivalence was the novel phenomenon to be explained, not the well-known fact that large species are rarer than small ones (Mohr 1940). Granted, with suitable care these allometric relationships can guide the prediction of population density (Marshall et al. 2021), but that is not their significance here.

Energy equivalence only partly explains a species' population density. Density depends on factors in addition to nonadaptive selection within a trophic level, two of which are of notable importance. First, individuals of

A

Effect of metabolic level

B

Effect of trophic level

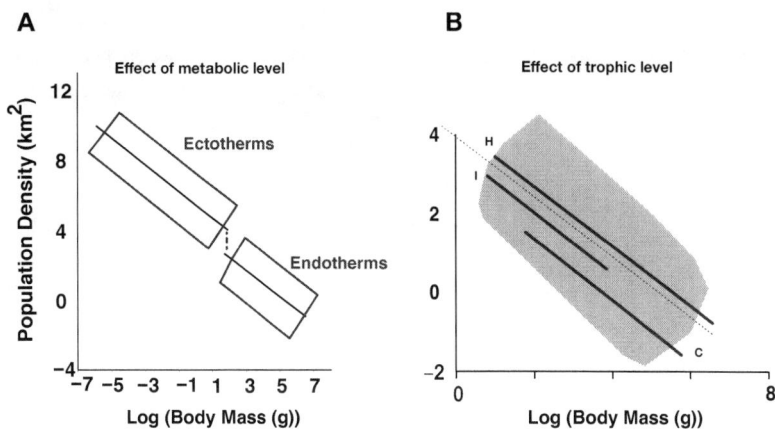

FIGURE 10.1. Geometric effect of metabolic level and trophic level on density/mass plots. A: Endotherms exhibit lower population densities because of significantly higher individual metabolic requirements. B: Higher trophic levels are offset downward and typically exhibit different size ranges. Shown are three schematic trophic groups within the nonviolent terrestrial mammal data (shaded polygon). Herbivores (primary consumers) are the highest, invertivores are lower, and carnivores that as a level feed on both H and I are lower still. Size range spanned by each group is indicated by its respective line. Dotted line is overall line. For clarity, lines drawn are all −0.75 in slope, and minor differences in empirical slopes have been ignored. See Damuth (1987).

different species may differ in their energy use because of physiological differences unrelated to overall metabolic scaling (e.g., ectotherms versus endotherms). Individuals of endothermic species use energy at approximately 25–30 times the rate that ectothermic individuals do (Hemmingsen 1960), so when their populations are using the same amount of energy as ectothermic populations, their population density is correspondingly lower (Damuth 1987; Gillooly et al. 2001). Endothermic species (birds and mammals) are concentrated at the high end of the body mass range. Second, the total energy in a trophic level depends on its rank in the trophic pyramid, because of the limited transfer efficiency of energy across trophic levels (Elton 1927). Thus species in higher trophic levels tend to have lower population density for the same body mass (Damuth 1987; Hechinger et al. 2011). Trophic levels often exhibit different body size ranges. For example, predators are usually much larger than their prey (see section 7.5). The typical effects of these factors are illustrated in figure 10.1.

To the extent that these factors vary with body mass or are restricted to certain ranges of body mass, they not only contribute noise to the size-density plots but may also affect the geometry of the plot. This in turn will produce deviations in fitted statistics that do not straightforwardly reflect relative energy use within trophic levels. If we want to examine density scaling of all the species together, as if we were comparing physiologically similar species within a trophic level, we should do our best to remove or minimize the effects of these factors, where present.

For example, imagine that we would like to compare animal population energy use, but we have density instead for a mixture of endothermic and ectothermic species. Furthermore, endothermic species are concentrated at large body mass. We ordinarily can convert density to energy use by multiplying density times metabolic rate, but the metabolic rate for endotherms is 30 times higher than it is for ectotherms. That means the same amount of energy will support 1/30 the number of endothermic individuals of a given mass as ectotherms. This situation is depicted in figure 10.1A: ectotherm and endotherm densities are offset to reflect the factor of 1/30, although species populations within both groups actually use similar amounts of energy. Each group by itself exhibits identical slopes and energy equivalence, and overall energy equivalence among all species holds. However, fitting a line to the raw density data as a whole (pooling ectotherms and endotherms) would generate a steeper slope than would be expected for energy equivalence, leading to the incorrect conclusion that energy equivalence overall is not supported. To use all of the species together, we would want each ectothermic individual to be comparable to the same number of homeothermic individuals; the scaling among all of these species taken together could then indicate whether energy equivalence was supported. In effect, standardizing metabolic level would make it unimportant for the analysis whether a species is endothermic or ectothermic. Metabolic level can be corrected for endothermic/ectothermic differences directly (Damuth 1987) or indirectly (Gillooly et al. 2001; Hechinger et al. 2011).

Likewise, size-density studies that include multiple trophic levels often identify many effects of relationships *among* trophic levels that would obscure patterns within trophic levels if all species were analyzed together (Cohen, Jonsson, and Carpenter 2003; Jonsson, Cohen, and Carpenter 2005; Rossberg et al. 2008). Inspecting figure 10.1B, it appears the best strategy in the present case would be either to study only one trophic level at a time, or if necessary to correct the data for the trophic level offset(s) if the ranges of body mass of the trophic levels differ greatly (Hechinger et al. 2011).

In figure 2.5, data on individual communities (A–C) were corrected for temperature and for offset due to trophic level membership. Thus each represents an approximation of a single trophic level. Data for the global construct D were corrected for endotherm/ectotherm offset, but trophic levels have not been standardized. The somewhat greater spread of points in the community regressions as opposed to the global construct may be due to the construct values being species averages over multiple communities as opposed to observations of individual populations (cf., Silva and Downing 1995).

10.3.2 Why Ecological Allometries Are Often Misunderstood

Ecological and life history allometries have frequently been misunderstood, and that confusion has its roots in two tendencies of thought among biologists, both artifacts deriving from the history of the term. During their training, most biologists first encounter *allometry* in the original context of relative growth of body parts (Huxley 1932; Gould 1966; see also Thompson 1917). Before the Modern Synthesis, many suspected that Darwinian natural selection could not explain patterns of shape change in lineages (Stevens 2009). An allometry was conceived of as an empirical, strong functional or developmental (and later, genetic) constraint that linked characters in a particular way, such that selection on one character necessarily caused a disproportionate change in another. Concern was expressed that under adaptive natural selection involving size, such allometries would necessarily constrain or drive evolutionary trends in correlated trait values, creating monstrosities in the process and impairing adaptedness, or even leading to lineage extinction. However, as the synthesis progressed, functional and developmental allometries ceased to be seen as a major problem for lineages.[1]

What remained was the idea that biological allometries were rigid relationships among characters, whose origin was often not known. Each one was a given, presumably a historical artifact of linkages caused by the complexities of development or other processes. Significantly, to say that a group of species were subject to or following a particular allometry was to imply that all other things being equal, the species' characters relative to size would lie exactly on the observed allometric line. *Deviations* from the line represented adaptive evolution in each species to break the allometry. Selection was thus the explanation for the *spread* of points about the line, but not for the existence of the line itself; selection was not involved in either the origin or maintenance of the allometry (cf. a similar point made by Kauffman 1993, 15). The deviation from the allometric line could (with caution) be used as a "criterion of subtraction" for distinguishing adaptive evolution from the variation that was demanded by a change in size (Gould 1975; Smith 1984). This traditional view is often reasonable when dealing with developmental and functional allometries (e.g., Marcy et al. 2020), but in the case of ecological allometries that are better explained by nonadaptive selection, the traditional view can lead ultimately to paradoxical conclusions at variance with a straightforward interpretation of the data.

The first misunderstanding is the incorrect assumption that merely by describing a widespread interspecific scaling relationship across size, one is making the claim that there is some force that is "pushing" or "pulling" things to be *on* the best-fit line—the claim that somehow, if nothing else were happening but this one postulated force, all species would line up on that line. That implies that deviations are the result of some forces opposing the force pulling the species onto the line. Furthermore, this view also implies that the distance that species are away from the line represents the strength and nature of the opposing, special circumstance(s)—adaptations—counteracting the supposed centripetal force.

Second, if we assume that there are such evolutionary forces causing deviations, then it is reasonable also to assume that the more uniform the group of species analyzed, the more similar they will be in their "offline" special adaptations. Therefore, among themselves they should be *more* closely aligned to an allometric line rather than being as widely dispersed as is a less homogeneous sample of species. If the all-mammal line has a certain cloud of variability around it, then, for example, heteromyid rodents or deer species, taken alone, should show a smaller range of variability and thus a better fit to an allometric line.

Both of these standard assumptions are seen to be wrong in a case like that of the energy equivalence relationship. When interspecific allometric relationships started being described in ecology (McNab 1963; Damuth 1981a, b; Calder 1984), there was no equivalent to the presumed absolute constraint of development. Allometries without an obvious mechanism to maintain them seemed suspicious. All things being equal, why would species line up on these lines? Adaptation presumably explained the deviations, but from what kind of constraint?

We believe that ecological allometries, as well as many others, can be explained more clearly as being the *result* of selection—nonadaptive selection on extreme deviants. Adaptive evolution does largely explain the deviations in each case, but nothing is pushing the species toward the line. For energy equivalence, there really is no known mechanism that can, all other things being equal, force species to apportion energy perfectly equally and thus force them to the line. The manifest lack of such a force causes some people to object that the data must not be indicating an allometric relationship in the first place, because of the incorrect assumption that allometry implies such a force. The second assumption is based on the first and explains why people expect to see conformity to the overall pattern even in samples of phylogenetically or ecologically similar species that differ hardly at all in body size (Bohlin et al. 1994). Failure to find this, and instead finding that variance within such groups is often as great as the overall relationship for a restricted range of body sizes, further cements the idea that there is something artifactual or wrong with the allometric relationship, because it casts further doubt on the efficacy of the (nondemonstrable) force pulling species to the line.

However, from the perspective of nonadaptive selection there is no mystery about what is happening. There is no force pushing things to be on the line (energy, density, or any other); phylogenetically closely related species are not constrained to be similar to each other in energy use and are free to vary at a given body size over the whole range available within the permitted envelope. Instead, species that are too far from the (energy use) line are unfeasible or go extinct. Nonadaptive selection results in an envelope within which considerable variation is allowed in the region of the line, but as extreme values appear they are continuously eliminated. With respect to the nonadaptive selection boundaries, all species are playing with the same criteria, though they may be using all kinds of different tools to jockey for position within the cloud. But there is no net, permanent success for any strategy, as the Red Queen knows.

Thus, nonadaptive selection offers a potential explanation for some ecological allometries, based on a different view from a common one of what it means to conform to an allometric relationship.

10.4 The Price Equation

The Price equation is a highly general description of evolutionary change that, among other things, provides a general definition of what it means for something to be a selection mechanism (Price 1970, 1972).[2] That is, the

Price formulation comprises a variational theory of evolution, in Fracchia and Lewontin's (1999) sense discussed in chapter 3. But what is more, the full equation (Price 1972) describes the entire process of evolutionary change in a given situation, including any transformative components, by making a "clean" distinction between change caused by selection alone and that caused by all other processes. We aim for a brief overview in this appendix; detailed, accessible derivations of the Price equation are available in many publications (Frank 1995; Rice 2004; Okasha 2006; Gardner 2008).

The Price equation describes change in the mean value of a character, z (which may be any kind of phenotype), from one time to another. It must be possible to assign relative fitnesses to the individuals (or types of individuals), based on the value of z. That is, individuals differing in z in the original ("ancestral") population have different expectations of representation in the next (future, "descendant") population, and these expectations are expressed by fitness values assignable on the basis of z. Frequently the biological interpretation of the Price equation considers the members of the ancestral population to be parent individuals and the members of the descendant population to be their offspring in the next generation. However, this is not required by the expression. The timescale of the before-and-after population comparison is not defined, except to the extent that it is the scale over which the fitness values have meaning. If selection is among entities that differ in stability or persistence but do not reproduce, then the "parent" and its "descendant" may even be the same entity observed at both time instances. *Representation* in the next generation is thus meant in the most general sense and may occur through any combination of differential survival or reproduction; the important thing is that there be a mapping from individuals in the first population to those "descended" from them in the second population.[3]

The total change ($\Delta \bar{z}$) in a population's mean value of the character (z) from one generation to the next is given by the Price equation:

$$\Delta \bar{z} = \text{Cov}\left(\frac{w}{\bar{w}}, z\right) + E\left[\left(\frac{w}{\bar{w}}\right)\Delta z\right],$$

where $w/\bar{w}$ is relative fitness.

The first term on the right-hand side is the covariance of the character with relative fitness.[4] This represents the change in $\bar{z}$ due to selection alone.[5] The covariance term defines a selection process and describes its operation over one generation. Such a process is extremely general, inasmuch as the only necessary condition is that the entities exhibiting z can be associated with fitness values. If z does not covary with fitness, then the covariance term will be 0 and there will be no selection; with respect to z, all individuals in the parental population would have an equal chance of representation in the descendant population.

The second term is a fitness-weighted average of the mean change, due to processes *other* than selection, in the phenotype of each of the elements of the parental population—when that element is projected into the descendant population. It is weighted by fitness of the parent, since parental individuals of low fitness will have lesser representation, and their transmitted changes will have a correspondingly lesser effect on the composition of the descendant population, and vice versa. Although this expectation term is often associated with processes of transmission of character values from parent to offspring (inheritance), it actually represents as well any change that would affect the phenotype of an entity during the time period in question, and that would be reflected in the associated individual(s) mapped from it into the descendant population. This can include a variety of biasing factors (Simpson 2010; Edelaar, Otsuka, and Luque 2022). Significantly, this term also includes *selective* processes that may be operating at a lower level, that is, *within* the individuals bearing z (e.g., segregation distortion within organisms). Internal, lower-level selection of this kind thus generally appears in Price formulations as a transformative process. If parents perfectly resemble their offspring in z, then the expectation term will be 0, and only the effects of selection (if any) will be manifest. If, at the other extreme, there is no statistical relationship between parent and offspring z values, the descendant population will just reconstitute the distribution of z found in the parental population; the covariance term and the expectation term will be of the same magnitude and opposite in sign. Selection will still be occurring, but it will have no effect on $\bar{z}$.

The Price equation describes the effect of selection but does not require biological details characteristic of organisms, such as genes, development, or reproduction. Imagine that the entities undergoing selection are ones that are not considered to reproduce as such (e.g., most physical entities, or parts of multispecies interaction networks, such as the three-species modules described in chapter 2); further suppose that selection acts only on some property of the system, z, that confers stability and thus longevity. We would expect the covariance term of the Price equation to represent change in the frequency of individuals with different values of z (different modules) over time. However, it may also be that these entities are composed of subunits (such as populations of individual species) that can change in nonrandom ways that, for whatever reason, may affect the value of z over the same stretch of time. The Price equation gives us a framework to distinguish evolutionary change in z due to selection on persistence from changes in z due to other processes that may be happening simultaneously. The net result of these different influences on change in z will depend on the relative magnitudes and signs of the two terms of the Price equation. Application of the Price formulation to the evolution of triplet modules does not violate any tenets of evolutionary biology, since it deals only with the selection process among specific entities and makes no assumptions about other biological processes that may be occurring, whose effects are represented as transformational. Thus, in keeping with a growing consensus, the Price equation is applicable to any population undergoing a selection process, and this includes macroscale multispecies systems and their multispecies components.

Because of its generality, the Price equation in the form just presented provides only limited insight into many evolutionary subjects. The real power of Price's formulation comes from the fact that the terms can be partitioned in many ways, adding assumptions and specifics and isolating different aspects of evolutionary change. Variances, covariances, and regression coefficients form critical building blocks of evolutionary theory (Rice 2004), and their algebraic relationships allow expressions ultimately derivable from the Price equation to address a wide range of questions, and even to recover results from research traditions whose theories were not based originally on any explicit connection to the Price equation (Hamilton 1975; Wade 1985). In fact, Queller (2017) argues that the Price equation should be regarded as *the* fundamental theorem of evolution, from which other fundamental relationships, theorems, and representations of selective evolutionary processes can be derived (e.g., Fisher's fundamental theorem, the breeder's equation, multivariate selection, Hamilton's Rule/inclusive fitness approaches, multilevel selection). The Price equation provides a means of conceptual unification of selection studies throughout evolutionary theory and even to selective processes outside of biology (Frank 1995; Luque 2017; Lehtonen 2018).[6]

Finally, the Price equation, in describing the mechanics of change under selection, tells us nothing directly about adaptation. The fact that the Price equation applies equally to selective processes in general—as Price himself believed and is widely accepted—including those that occur where adaptation is impossible, implausible, or ill-defined, emphasizes the fact that selection and adaptation are distinct, and that a selection process without adaptation is fundamentally no different from a selection process in which adaptation is occurring.

10.5 Kepler's Planetary Allometry

An example from outside of biology provides a clear and well-understood illustration and also suggests how natural it is to apply a version of non-adaptive selection to explain long-term regularities observed anywhere in nature. The case of the 400-year-old laws of planetary motion contains all of the logical elements of the theories that we construct to account for ecological regularities. The structure, consisting of primary mechanisms, a selective process, and resultant patterns (in this case usually called laws) is clearly discernible when one considers the relationship between the observed properties of the solar system and their explanation via physical law. In the case of Kepler's third law, the pattern is also a power-law relationship (see appendix 10.1), similar to the ecological allometries discussed in chapters 2 and 5.

Johannes Kepler (1571–1630) discovered a set of three regularities in the motion of the planets that are known as Kepler's laws:

1. The orbit of each planet is an ellipse with the sun at one of the foci.
2. Each planet orbits the sun such that the radius vector connecting the planet and the sun sweeps out equal areas in equal times.
3. The squares of the periods of any two planets are proportional to the cubes of their mean distances from the sun (i.e., a kind of planetary allometry).

For Kepler, these were empirical regularities and were not predictions from theory. Kepler's laws were fully explained a generation later by Newton's law of gravitation. This much is well known, but an important additional fact that is usually not mentioned is that our current solar system, which Kepler was observing, is highly selected: the original set of myriad interacting bodies forming the protoplanetary disk did not conform to Kepler's laws. Three or more strongly interacting bodies are unstable under gravitational laws and do not generally exhibit regular and repeatable motions. Thus, were Kepler to have observed the early solar system, it would have appeared chaotic, and there would have been no simple regularities to record. It is only now, when the system has evolved to a stable set of planets that effectively do not interact with each other, that we can view each planet singly along with the sun as a two-body problem. This allows us to obtain Kepler's laws as a simple consequence of the Newtonian law of gravitation. On the way to a stable configuration, there was a long history of bodies falling into the sun or colliding with each other—a process of eliminating unstable (more-than-two-body) configurations that has resulted in the evolution of the stable solar system that we now observe.

That is, we have a bounded, distinct population of particles of varying sizes which, through a process of aggregation, become large enough to form planets that are found at various distances from the star. At that point they interact gravitationally primarily with the star, and the configuration becomes progressively stable.

Thus we have the components of a nonadaptive selection mechanism that explains the observed motion of the planets:

1. This primary cause or physical mechanism explains planetary motion: Newton's law of gravitation.
2. This mechanism gives rise to a *selection process* that acts to eliminate unstable configurations (when present), leaving a system comprising only stable components.
3. The resulting stable configuration of the evolved planets whose motion under gravitation conforms to Kepler's laws, which are empirical and persistent patterns, and that arise as a result of the primary causal process and the selection process it creates.

Further insight comes from considering how Kepler's third law appears as an empirical finding (see figure 10.2). The resultant pattern that emerges from the causal mechanism and the selection process is a power law, reminiscent of the allometries frequently seen in ecological research. Knowing only the power law, we would be tempted to regard conformity to this relationship as an intrinsic, individual (but otherwise unexplained) property of each planet, taken alone. Seeing the entire theory, we recognize that conformity to the power law arises and is maintained as the stable end state of an evolved system.

The selective process derives from the law of gravitation, in the sense that three or more interacting bodies are always unstable; progressive reduction of the system to objects that interact strongly with only one other

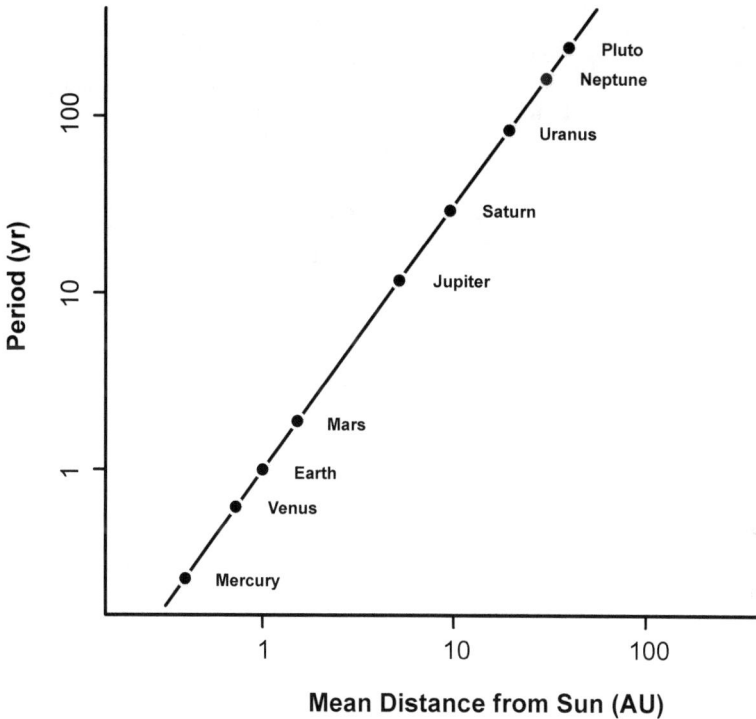

FIGURE 10.2. Relationship among planets in Earth's solar system between the length of time that it takes to revolve around the sun and the mean distance from the sun. AU represents astronomical units, which is the average distance from Earth to the sun. The y-axis is the period of rotation around the sun, in years. Both axes are log-transformed.

object stabilizes it. The process in this case is entirely one of elimination and amalgamation, as no new material is being produced. As such, the selection process and its ancillary properties bear only a family resemblance to Darwinian natural selection, and we do not think it can reasonably be considered to be a version of Darwinian adaptive selection acting at the planetary scale. This difference from Darwinian selection is not a flaw or limitation; one is not trying to answer the question of whether the solar system or the planets are adapted to local conditions, or whether they will change their mean properties if the environment of the solar system changes. In no sense is the solar system "better," "optimized," or somehow well adapted to conditions after the long exposure to selection, and in this sense the selection process is clearly nonadaptive. The selection process nevertheless acts effectively to generate Kepler's laws, and it is an entirely natural process of stability selection that does the work of translating from short-timescale mechanism (Newton's laws) to the long-standing pattern (Kepler's laws) in this particular theory. Similar stability selection processes, acting on other spatial scales, are proposed by astronomers to underlie the evolution of planetary rings and of empty spaces in the asteroid belt (Cuzzi et al. 2010).

Recent observations of planets and protoplanetary disks around distant stars fully confirm the view that our planetary system evolved out of the material generated by the sun. As we now know, planetary systems are found around most stars (Cassan et al. 2012). These systems differ considerably in characteristics such as the numbers and sizes of planets and their distances from the star. But it is most unlikely that the differences among planetary systems of different stars are the result of adaptive evolution to different "environments," or to evolved conditions (Winn and Fabrycky 2015). What is more likely is that the diversity represents chance variation in the operation of a single kind of nonadaptive selection mechanism (and this is how it is interpreted by astrophysicists).

10.6 Selection Analyses, Multilevel Selection, and Adaptation

Over the last 50 years there has been vigorous debate about the significance of selection and adaptation in structured populations (metapopulations), that is, populations that are divided into permanent or temporary subpopulations, social groups, family groups, and the like, but are united by some degree of migration among the subunits. Arguably this is a more realistic description of many populations in nature than is the single well-mixed population of classic natural selection. Ecological heterogeneity and extinction/colonization dynamics in metapopulations affect the outcome of both ecological and evolutionary processes in many ways (Hanski and Gilpin 1997). But what is relevant to us here is a particular set of contentious issues that have been extensively discussed under the general topic of "units" or "levels" of selection (also widely contrasted in terms of the purported differences between the approaches of *kin selection* and *group selection*). The complex, intricate, and at times acrimonious history of this subject is beyond the scope of this book (recent reviews include Sober and Wilson 1998; Lloyd 2001, 2012; Okasha 2006; Wade 2016). However, there is a specific issue involving group-structured populations that further illustrates the distinction between adaptation and selection we have been discussing and also makes it clear that we are not proposing a multilevel selection scenario for nonadaptive selection.

In hierarchically structured populations, the fitness values of individuals depend not on just the individuals' interactions with a single, shared environment, but (also) potentially include individuals' interactions with a wide variety of factors owing to the individual's membership in one or more "groups," broadly defined (Wade 2016). Groups in this sense describe any features of population structure that result in consistent non-randomness in the interactions of individuals. If group membership affects individual fitness, then population structure in the broadest sense can possibly lead to evolutionary outcomes not possible under mass selection (e.g., the evolution of altruistic behavior, in which individuals experience a fitness cost in the process of enhancing other individuals' fitnesses). One of those group-related factors may be interactions with the other individuals that the group comprises. Another (not necessarily distinct) factor may be the fates (longevity and proliferation) of groups themselves. Significantly, groups need not exhibit complex adaptations, high levels of organization, or organism-like behaviors; if groups merely facilitate non-random interactions among individuals that affect individual fitness, then the effects of population structure may be seen (Wilson 1980).

The dynamics of selection in hierarchical populations have historically been addressed by two research traditions that at first seemed to model selection very differently—*multilevel selection*, in which the selective forces within and among groups are explicitly modeled (Wade 1978, 1985; Wilson 1980), and *inclusive fitness* theory (Hamilton 1964; Maynard Smith 1964; Taylor and Frank 1996), in which fitnesses are evaluated and combined across population structure to produce an overall fitness for individuals or alleles. Over most of the history of the subject, it was generally thought that the different ways these traditions represent selection dynamics represented different causal processes and hence had different implications for adaptation. Now it is increasingly recognized that the basic selection models of these two traditions are mathematically equivalent (Hamilton 1975; Wade 1980, 2016; Traulsen 2009; Marshall 2011).[7] Thus, they must be describing the same selection phenomena. Each of the two versions of a general selection analysis is suited for different goals (Goodnight 2013).

However, the modern controversy has from the beginning been about adaptation, and (now we realize) not about selection models (Williams 1966; Brandon 1978; Wilson 1983). Elizabeth Lloyd (1988, 1994, 2001, 2008) persuasively argued that the conflation of group selection and group adaptation has bedeviled the discussion from the start. Researchers in the levels of selection debate disagree less over how selection is working than they do over the way the historical pattern of selection can inform one about what counts as an adaptation and, in particular, who or what "benefits" from the adaptation and how evident the adaptation must appear as an engineering achievement of a history of selection, specific to a particular level of biological organization. In the context of multilevel selection, the issue is often conceived as whether groups can undergo selection and adaptation "at their level" and exhibit group adaptations, or whether everything must be considered a form of individual selection "in context" (Gardner and Grafen 2009; Winther, Wade, and Dimond 2013).

The levels-of-selection literature reveals a long struggle with how to reconcile patterns of selection with the process of adaptation (see Okasha and Paternotte 2012; Okasha 2016). Repeatedly, researchers have been tempted to try to deduce the process of adaptation from the pattern of selection by itself (Sober and Wilson 1998; Gardner and Grafen 2009; Sober 2011; see also Wimsatt 1980; Wade and Kalisz 1990; Wade et al. 2010; Wild, Gardner, and West 2010; Shelton and Michod 2014b). This is a different issue than the historical and nonhistorical usages of adaptation described in the text of this book. Here, the issue is an attempt to identify and locate the adaptations from the pattern of selection alone.

Such attempts largely fail to resolve issues about adaptation. We think that the obvious conclusion is that an analysis of selection dynamics is not by itself an analysis of causes (cf. Allen and Starr 1982, 52–53; Heisler and Damuth 1987; Wade and Kalisz 1990). The best that can be done using selection dynamics alone is to conclude that where there is selection, there is a *possibility* of adaptation occurring at a particular level (Shelton and Michod 2014b). But as we have argued throughout, verifying that adaptation really is occurring and identifying the adaptation itself requires different information and a causal argument outside of the selection dynamics themselves, and this is no different in a structured population situation than it is in a classic, undivided population (see Birch 2019 for a similar perspective on the place of causality).

An example may help to clarify the point. Suppose that there is a prey species that lives in fairly isolated groups, and among the individuals there is variation in conspicuousness to their primary, visual predator. There is variation among groups in the proportion of conspicuous and inconspicuous individuals. A predator individual forages throughout the environment until it detects a prey group. (It detects the group as a whole visually at some distance, rather than bumping into a prey individual.) Then the predator proceeds to the detected group and commences feeding, eating whatever fraction of the individuals it can catch, but is more likely to catch and consume the more conspicuous individuals. This can be described as a multilevel selection scenario: The probability that your group will be attacked depends on its *phenotype* (the frequency or number of conspicuous individuals), and once your group is found, you are selected against if your phenotype is conspicuous. In this example, then, we know the real causes of fitness.

We could model the selection in this case either as multilevel selection or from an inclusive fitness perspective, and we would necessarily get the same result. It is not difficult to predict that the ultimate outcome will be the disappearance of the conspicuous individuals. From the point of view of multilevel selection, individual selection (predation probability within groups) and group selection (detection of groups) are operating in the same direction. A contextual analysis (Heisler and Damuth 1987; Goodnight, Schwartz, and Stevens 1992; Okasha 2006) would show that there is a group effect on individual fitness (both individual and group selection are occurring). But in a mathematically equivalent inclusive fitness treatment, it is equally and certainly true that the selection against conspicuous individuals will represent an optimization of inclusive fitness, and the conventional interpretation would be that only individual selection is occurring. However, neither analysis alone would demonstrate *why* selection is occurring, or what the relevant adaptation(s) being formed might be.

But for this example we *know* why. From the point of view of what problems were solved (and thus what the adaptation[s] might be), it is difficult to avoid the interpretation that there seem to be two separate problems "posed" by the environment, each relevant to a different level, and both are ameliorated by the joint selection against conspicuousness. One is a problem for the group: the probability of detection of groups by the predator. The other is avoiding being eaten once your group is found, which is an individual problem. Note that (unlike in the example) solutions to these two problems could be related to variation in distinct and unrelated traits. Furthermore, the situation is unchanged if the groups exist only temporarily during one stage of the life cycle or seasonally.

How one decides what this knowledge of the true causality reveals about adaptation is a little unclear. Is it two characters (a group one and an individual one) and two adaptations (a group one and an individual one)? Is it one character (minimal conspicuousness) that contributes simultaneously to two adaptations? Is it one adaptation that somehow straddles both levels? The fact that the answer seems not to be informative once the causality is understood suggests that trying to locate the level(s) of the adaptation(s) may be the wrong question to ask of a selection analysis, and we will not pursue it further. Our point is that in any case, the pattern of selection alone cannot resolve it.[8]

Another example provides an instructive contrast. Imagine a species in which the groups offer only an interaction venue for individuals to work out their relative fitnesses, in the context of the interactions among group members. As mentioned previously, this situation can generate group effects on individual fitness, yet one would be hard pressed to claim that the group was "solving a problem" at its level or developing a group-level adaptation. We don't award gold medals to the Olympic stadium. Yet the dynamics of that multilevel selection would look similar however they are modeled; from the selection analysis alone one could tell the direction of change of traits but would not know the level(s) at which adaptation was taking place. Nothing could be a clearer demonstration that the pattern of selection does not adequately and meaningfully capture the idea of adaptation, especially across levels of organization. One needs to go outside of the dynamics and investigate the causes with other methods.

10.7 Selection without Adaptation: A 2,300-Year History

A natural process of selection without local adaptation is not a new idea; selective removal of the unfeasible or unstable enjoys a long history in Western thought. The pre-Socratic philosopher Empedocles (ca. 493–433 BCE) is usually cited for the earliest description of something akin to organismic natural selection (Zirkle 1941; Roux 2005). In his poetic vision, he imagined a time when, in an episode of prodigious creation, the earth produced the body parts of all biological creatures. These then combined haphazardly to form all manner of organisms—many, if not most, bizarre and monstrous (such as oxen with the faces of humans). Only those species that were by chance "whole-natured" and had parts well suited to each other were able to survive and reproduce, and the rest perished. Thus, for Empedocles, selection against the unstable or unfeasible primarily explained the harmonious combination of the parts of a functioning organism, without need for a designer or a teleological principle.

Empedocles's selection process was adopted by what became one of the major classical schools of thought, the materialistic philosophical sys-

tem of Epicurus (341–270 BCE). Epicureanism supported a purely naturalistic origin for the diversity of existing species. The Epicurean Lucretius
(99 BCE–55 CE) gives an account of the origin and fate of species that
is derived ultimately from Empedocles but omits the more fantastic elements. Gone is the independent creation of morphological fragments that
wander about in search of a body; in the remote past, the earth spontaneously created whole, functioning species. Moreover, chimeric mon-
sters and fantastic beings such as centaurs or human-faced oxen were not
produced; they were entirely unfeasible to begin with. Nevertheless, the
apparently well-designed species that we see around us derived from a selection process among a wide range of other species, all of which had been
produced by unspecified generative processes of the early earth. Some
of those species were barely feasible or generally inept and have been
lost to the modern world. For Lucretius, there is no reason that this selection process is not still active today, but the earth is now "worn out" and
incapable of producing new species (Lucretius 2001, bk. 5, lines 826–36).
In present times, species simply reproduce themselves. Thus, the selection
process was seen as a unidirectional one that has largely come to completion (similar to the evolution of the solar system mentioned in appendix 10.5). Species themselves were apparently regarded as immutable; selection was only an elimination process and did not create new diversity,
nor was it responsible for increasing the degree of adaptation of any given
species. That is to say, there was no process of adaptation that could create
novel traits; indeed, Lucretius appears to have regarded this as a logical
impossibility (along with teleological or final-cause explanations in general) (Amundson 1996; Lucretius 2001, bk. 4, lines 823–58).

In contrast, the other major classical Greek philosophical schools
supported divine or teleological sources for the apparent design of existing species and rejected the Epicureans' purely material mechanisms
(Roux 2005). Aristotle addressed the issue directly and at some length in
his *Physics* (Aristotle 1930). He rejected the general absence of purpose
in nature and could not accept the idea that species had ever been created with random properties. For Aristotle, the manifest design of organisms and the ways in which species apparently serve each other pointed
to a purpose behind the existence of each species and the development
of their physical form. The effective agent of development or "seed" of
each species generates its intended adult morphology, which is identical
for all individuals of a species (since the morphology fulfills the species'
purpose). However, Aristotle did not reject the idea of selection among
variants produced by chance: "Hence clearly mistakes are possible in the

operations of nature also. If then in art there are cases in which what is rightly produced serves a purpose, and if where mistakes occur there was a purpose in what was attempted, only it was not attained, so must it be also in natural products, and monstrosities will be failures in the purposive effort. Thus in the original combinations the '[man-faced] ox-progeny' if they failed to reach a determinate end must have arisen through the corruption of some principle corresponding to what is now the seed" (Aristotle 1930). For Aristotle, such selection is always occurring, even in the present day, and simply acts to preserve what later biologists would refer to as the species type.

This view that organisms and species originated from and represent the work of a teleological principle or a designer, and that eliminative selection among organisms serves only to maintain the integrity of this unchanging type, had substantial appeal to both what emerged as the dominant schools of classical philosophy and, later, Christian thought. What we would now call adaptation to local conditions did not demand a naturalistic cause and in fact was regarded as evidence for divine purpose. Species were fixed, and their fit to environmental conditions was a feature of their creation, not something that resulted from a process of transformation. Throughout ancient and medieval philosophy, the separate components of Darwinian evolution by natural selection can be occasionally discerned, but they are nowhere put together (Zirkle 1941). The immediate predecessors of Darwin, even those who accepted the mutability of species, still saw selection in nature, if important, as a process of which the sole effect was to contribute to the fixity of species (Gould 2002, 137). Pre-Darwinian evolutionists had a variety of other theories of species transformation, but they did not connect selection to adaptation (Corsi 1978, 2005).

So selection without adaptation has long been part of the intellectual landscape. Scholars have had no difficulty believing that selective processes affect the representation of biological forms, whether or not the scholars were concerned with local adaptation or even believed in evolutionary change. From Aristotle to Darwin, the selective removal of organisms that were relatively unstable, unfeasible, or unfit for local conditions had been part of scholarly thought. However, the full implications of locality-specific selective processes had not been understood.

Darwin and Wallace realized that evolution by natural selection could, under common circumstances, also explain the local adaptation of organisms and thereby also the enormous diversity of life. The power of this insight is difficult to overstate; it reverses the emphasis on natural selection from processes that constrain or counteract variation and diversity to an emphasis on local selection as the primary cause of the diversification of organisms, adapting them to specific, diverse, and ever-changing local conditions. Since the triumph of the neo-Darwinian synthesis, selection in nature has been closely associated with the Darwinian process of local adaptation. Organismic adaptation is a primary focus of evolutionary studies, simply because organisms and their populations possess qualities (such as individuality, heritability, and reproductive excess) that make them easily susceptible to Darwinian adaptive natural selection. However, that association does not preclude the existence of other selective processes, which do not lead to local adaptation. Some nonorganismal types of biological systems, especially macroscale multispecies interaction networks, faunas, and taxa, may not (easily) adapt, as wholes, to local conditions. Over the last 150 years people have understandably been deeply interested in making sure that we have a good explanation for the diverse characteristics of organisms and their fit to local environments. But it is also of importance to explain repeated changes and regularities in the structural characteristics of multispecies interaction networks—which do not seem to be well explained as examples of adaptive evolution involving the systems as wholes.

10.8 Selection and Adaptation outside of Biology

The selection process found in Darwinian adaptive evolution is a special case of biological selection, and the Price equation and its derivatives (Queller 2017; see appendix 10.4) provide a clear, concise, and very general description of selection in biological contexts, whether or not adaptive evolution is occurring as a result of that selection. But it has been common for people to generalize "natural selection" to a broad range of contexts *outside* of biology, or at least outside of the context of organic evolution. Price himself thought that *selection* was generalizable to such contexts (see comments by Frank 1995; Price 1995, written circa 1971). Moreover, if selective processes are in fact involved, most people assume this means that "adaptation" is necessarily occurring (or could occur) as a result.

Certainly not all abiotic patterns and processes are examples of selection. For example, the erosional transformation of mountain ranges into sedimentary basins does not seem to be meaningfully considered a selection process (but see Van Valen 1989). Such long-term nonselective processes operate to produce long-term results, can be directional, and can even go to an apparently completed (and at least temporarily more stable) state. In such cases, the temporal extrapolation of the relevant short-term causal mechanisms is effective and occurs in an obvious way. The dynamics of such physical systems or phenomena are not well represented as a series of selection events. Such processes generally represent forms of *transformational* theories, in the terminology of Fracchia and Lewontin (1999).

Extensions of selection to nonbiological and nonstandard situations are subject to three major issues. First, does the "population" make any sense? As argued in section 10.13.1, selection can "work" on any defined population but means nothing if the population is not composed of entities whose relative "fitnesses" make sense with respect to each other. Second, the "phenotypes" and their "fitness" values themselves must make sense with respect to the "generation time" of the selection process. And finally, the requirements of shared environments and system memory for adaptive evolution have to be met if adaptive evolution or "progress" is to be claimed.

10.8.1 Universal Darwinism? Selection in the Physical Sciences Is Mostly Nonadaptive

Dawkins (1983) first used the term *universal Darwinism* to refer to the fact that biological natural selection was likely to be found throughout the universe. Subsequent authors have extended the usage to refer to many overarching schemes that claim natural selection as a basis for the evolution of "everything." Sometimes "natural selection" seems to be used as a metaphor for the engine of perceived historical progress in all of existence, or to claim that the evolved characteristics of humans constrain or limit the progress of human history and all human-mediated phenomena (e.g., Cziko 1995; Dennett 1995; Villmoare 2023; Wong et al. 2023). Others postulate selection among multiple universes (Maynard Smith and Szathmáry 1996; Smolin 1997; Gardner and Conlon 2013). We regard these schemes as either metaphorical or necessarily incomplete and speculative rather than studies of selection per se.

The physical sciences and the sciences of complexity have toyed with the idea that selection underlies regular behavior of natural physical systems, the origin of ordered states, and responses to external stimuli. In such schemes it can be difficult to distinguish transformative from variational (selective) processes. Simple nonliving systems in nature nevertheless often show selective but not adaptive behavior (e.g., Bernstein et al. 1983).

So are most selective processes in nature outside of biology nonadaptive? An argument can be made that they are. The requirements of adaptive evolution are onerous for most nonbiological situations. Excluding artificial systems and those systems that are too closely bound up with human intentions (history, culture, ideas), change by "natural" selection outside of biology has more of the character of nonadaptive processes as we describe them. Even if a given change is initiated by an environmental stimulus, physical selective processes tend to operate on elimination of unstable configurations and the proliferation or retention of long-lived ones, based on intrinsic system properties. Nonadaptive selection does not end up having changed those intrinsic properties when a new equilibrium is reached. No new "adaptations" are being built; no new properties are being evolved. There is no ecological "problem" that is being solved. The system is not a "better" one after the selection, and individual system components do not "benefit" from the selection. Recall how we offer the evolution of the solar system as an example of nonadaptive selection, but not adaptation (see appendix 10.5).

10.8.2 Social and Cultural Change

But what about "nonbiological" selection in human activities? Here we are close to straying outside of the scope of our book. Social systems and other man-made systems may "evolve" or progress through history, but there is no reason to believe that such changes are governed primarily by a set of selective processes that resemble Darwinian natural selection. They are essentially artificial products of human activity and intention. "The economy" has often been compared with biological evolution (Nelson and Winter 1982; Hodgson 2002; Cordes 2006; Vermeij and Leigh 2011; Eichholz 2014). But why can't humans have built, sometimes intentionally and sometimes by trial and error, an independent way of increasing productivity, diversity, and technological sophistication that is based partly on adaptive evolutionary principles? The economy has no independent way of functioning without direct human action.

A carefully constructed artificial system can, indeed, show aspects of adaptive selection, although the relationship to Darwinian natural selection is often metaphorical (Holland 1995). The two most widely known "Darwin-ian" schemes for social and cultural evolution are the "memes" of Dawkins (1976), which supposedly operate like Dawkins's genes (see also Dennett 1995), and the "double-inheritance" scheme (Boyd and Richerson 1985), which relies on "cultural inheritance." As Fracchia and Lewontin (2005) pointed out, both such schemes assert that the cultural side of things oper-ates like an inheritance system of self-replicating units of (cultural) evolution, but these assertions are just metaphors; selective mechanisms can be made to appear to "work," but that does not mean that they will explain anything.

One specific case of cultural evolution, that of the selective progress of scientific theories (a form of "evolutionary epistemology"), has intrigued philosophers and scientists (Popper 1972; Toulmin 1972; Hull 1988), though it is not immune to collapsing into metaphor. Here, at least, there seems to be a selective process in which the better ("fitter") theory will win out. But the resolution of theory conflict seems in practice anything but the simple evaluation of a theory's "fitness" (Latour 1987; Hull 1988).

These observations, and other perspectives, support the idea that de-velopment, not selection, may be a richer metaphor than selection for the evolution of culture, and that the transformational elements, rather than the variational elements, are of more explanatory value in approaching the topic (Caporael, Griesemer, and Wimsatt 2014; Love and Wimsatt 2019). Recognition of this would imply that any selective content of culture is of secondary significance and may be largely unimportant or neutral. Wide-spread adaptation seems unlikely, but nonadaptive selective processes, because of their generality, could be of some explanatory value.[9]

10.9 Theories, Laws, and Models as Contemporary Philosophy Sees Them

Let us consider what kinds of things make up the conceptual structure of a scientific field, according to philosophers of science. The traditional philo-sophical view of theories (which held sway in the early to mid-twentieth century) regarded a science as composed of a formalized logical structure in which statements about the real world were deduced from axioms, laws, and other statements that made up an overarching, ultimately monolithic, theory. Laws were regarded as universal and exceptionless generaliza-tions that expressed fundamental truths about the universe. This view is variously known, with different emphases, as the received, logical positiv-ist, logical empiricist, or syntactic view.

Since the 1970s, philosophers increasingly have regarded such logical constructions as, at best, inadequate to represent the complexity of the way that theories and other explanatory schemes formally relate to one another and are constructed and used by working scientists (Griesemer 1984; Lloyd 1988; Thompson 1989; Cartwright 1999; Giere 1999). Instead, in contemporary philosophy, a science is considered to be made up of a potentially large number of (depending on the author) *models, theories,* additional conceptual tools, information about the empirical world, and information about the way that these all fit together (Lloyd 2015; Winther 2021). The detailed relationships, and even identities, of these ingredients are not agreed upon, but we can discern a common theme. *Theories* (or theoretical models) — when included — are more abstract parts of the conceptual landscape and may be more or less "formalized" (e.g., expressed in mathematical terms).[10] Less abstract *models* (or chains of models) connect general theory to particular situations in the empirical world, and vice versa. Models include sets of statements as well as nonlinguistic objects such as diagrams and physical models (Griesemer 1990a, 1990b, 2004) and define connections between concrete instances and abstract theory.

The relationship/distinction between models and theories is also in dispute. In some views, theories are "hierarchical and intertwined families of models" and have no existence apart from them (Lloyd 2015); other views see models as "mediating" between theories and the empirical world (Morrison and Morgan 1999). Something (a model) has to make a bidirectional connection between theories and reality; the law or theory alone cannot be truly general (formalized) and at the same time guarantee that it applies to any specific cases. Likewise, the natural phenomenon carries no label that says it is explained by a particular theory. The model supposedly provides internally all the necessary connections between the two. In practice there is a lively discussion about what models are, who makes them and when, and whether they do their work alone.

In principle, the models used by scientists and those used by philosophers are the same. However, there is no guarantee of this in practice, nor that philosophers and scientists necessarily need or want to use exactly the same pathways among models to satisfy their respective interests (see Nersessian 1999; Griesemer 2013; Otto and Rosales 2020).

Regardless of the details, it is generally agreed that models are doing much of the work of scientific explanation and theory building.

It may continue to be of philosophical interest to ask to what degree scientific theories and laws can be formalized, and what such formalizations (if possible at all) should be based on. But among working scientists theories are usually constructed, understood, and used without regard to how they may be later reconstructed according to particular philosophical formalization schemes. For purposes of this book, our usage of "theory" and other conceptual structures does not depend on any specific concept of scientific theories; rather, we see our usage as being consistent with the general philosophical trend toward viewing theories/models as manifold, hierarchical, interacting conceptual schemes built by scientists to explain phenomena.

Nonadaptive selective explanations can be thought of as theories (or models) about how part of the world works, but this does not fully capture their shared characteristics—only the fact that they tend to be broad generalizations rather than detailed models of specific local processes. To go deeper we need to consider another, related concept—in what sense scientific explanations and lawlike statements can exhibit the necessity and universality that are supposed to characterize something called a "law" (see section 3.2).

10.10 Heterozygote Fitness Advantage

Feasible stable equilibria are maxima of the average fitness (fitness peaks). We base the proof on the convexity of the average fitness function, which is required for the stability of the equilibrium. If W_{ij} is the fitness of a genotype (i, j), the convex function of allele frequencies

$$\overline{W} = \sum_{i,j=1}^{n} W_{ij} p_i p_j$$

describes the average fitness of the population for any set of frequencies $p_i, \ldots, p_n$ (where n is the total number of alleles).

A convex function is always above a linear function that can be strung under it using the corner values $W_{11}, W_{22}, \ldots, W_{nn}$ at points $(1, 0, 0, \ldots, 0)$, $(0, 1, 0, \ldots, 0), (0, 0, 1, \ldots, 0), \ldots, (0, 0, 0, \ldots, 1)$. Therefore

$$\sum_{i=1}^{n} W_{ii} p_i < \overline{W} = \sum_{i,j=1}^{n} W_{ij} p_i p_j$$

and this is true for all frequencies. Specifically, it is true for the frequencies at the center of the space: $(1/n, 1/n, \ldots, 1/n)$. Substituting this into the inequality above, we have

$$\overline{W}_{ii} < \frac{1}{n^2} \sum_{i,j=1}^{n} W_{ij}$$

The left side is the average homozygote fitness. We can transform the right side as

$$\overline{W}_{ii} < \frac{1}{n^2} (n\overline{W}_{ii} + n(n-1)\overline{W}_{ij})$$

where

$$\overline{W}_{ij} < \frac{1}{n(n-1)} \sum_{i \neq j} W_{ij}$$

is the average heterozygote fitness. Canceling n and collecting terms, we have

$$\overline{W}_{ii} < \overline{W}_{ij}.$$

It turns out that it is true that weighted averages of the homozygotes and heterozygotes (weighted by their frequencies at the equilibrium point) also satisfy this inequality (Ginzburg 1979). So whether it is through sampling the population or direct arithmetic averaging, the heterozygotes' average fitness has to exceed that of the homozygotes.

10.11 Models of Single-Species Population Growth

We can illustrate the way that instability arises in population models using as an example a simple model of population growth with nonoverlapping generations (Ricker 1954), one of many possible population models that have been used in ecology (May 1974):

$$N_{t+1} = N_t \exp\left[r(1 - \frac{N}{K}) \right].$$

The equation simply states that the population (N) at the next time step (t) is equal to the current population size times e (the base of the natural logarithms) raised to a power whose value, in turn, depends on the intrinsic rate of population growth (r) and how close the current population is to the carrying capacity, K.

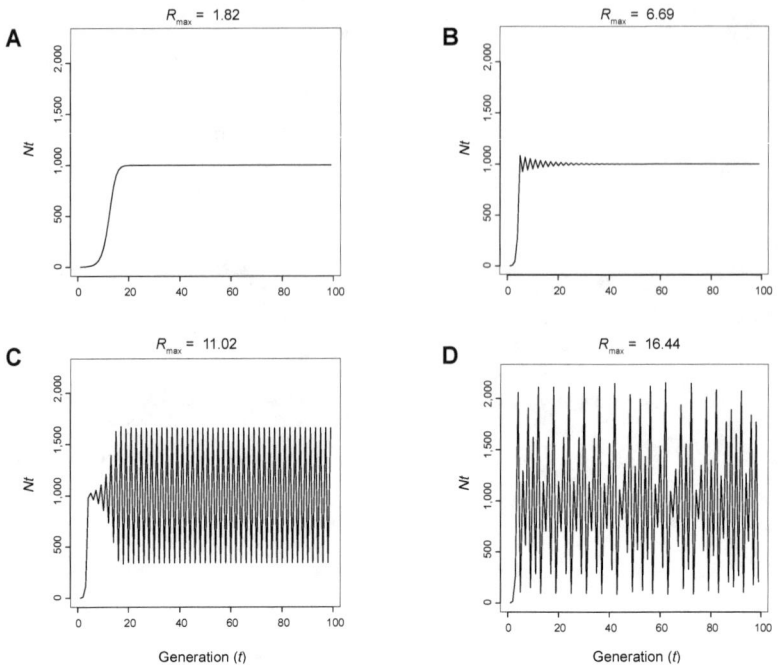

FIGURE 10.3. Population dynamics for different values of $R_{\max}$ in the single-species population growth model discussed in appendix 10.11

The parameter r represents the per capita rate of population growth over the interval t. In many treatments, t is a unit of calendar time. However, here we take the important step of changing timestep t to be one *generation* in length. The rate of increase r is thus the per capita rate that the population grows per generation. The model's behavior is unchanged by this change of scale. However, the change to generation time will allow us to compare population growth of species of different sizes, and thus different generation times, on a common, natural scale.

Following the above growth model, the population will start growing at low density at a rate per generation close to e^r. This is the value that we have defined in the text (see chapter 5) as $R_{\max}$. It is the net reproductive rate in the absence of density-dependent feedbacks. As N increases, population growth will then gradually slow down as the numbers of individuals begin to interfere with each other, consume more of the resources, and so forth—ultimately stopping growth when the population equals the carrying capacity K. Once at K the population will simply remain at that size unless perturbed by outside forces. This model can thus generate, over time, the familiar "S"-shaped curve of population growth, starting fast and gradually reaching equilibrium (figure 10.3A). However, this "nice" behavior characterizes only a range of low values of R_{max}. If R_{max} increases sufficiently, the population will overshoot K but exhibit damped oscillations that eventually return to a stable equilibrium (figure 10.3B). Further increasing R_{max} leads to stable but very large fluctuations (figure 10.3C) or, as R_{max} continues to increase, to chaotic, wildly fluctuating populations (figure 10.3D).

Figure 10.3 illustrates one simple model. Slightly different models of single-species *density-dependent* (i.e., N-dependent) population growth may be appropriate for different biological situations. But although the curves generated by different models of population growth will differ in detail, all such models share the property that there is a threshold value of R_{max} above which leads the population toward increasing instability and, ultimately, chaos.

10.12 Community Interaction Networks, Asymmetry, and QSS

In this book we have frequently used a general interaction matrix to investigate the stability of networks of ecologically interacting species. This depiction of the network takes the form of the classic community matrix, a square matrix (equal numbers of rows and columns) representing, for each species pair, the effect of the increase in abundance of one species on the growth rate of another (Levins 1968; May 1973a). For example, as we saw in chapter 2, a three-species network can be represented as

$$\begin{bmatrix} a_{11} & a_{12} & a_{13} \\ a_{21} & a_{22} & a_{23} \\ a_{31} & a_{32} & a_{33} \end{bmatrix}.$$

Here the interaction terms a_{ij} represent the effect of species j on species i; the diagonal terms represent a species' effect on itself (i.e., the effect of its density on its own growth rate). Interaction terms (a_{ij}) equal to zero indicate a lack of an effect, and otherwise interaction terms may be positive or negative (i.e., ordinarily the effect of an increase in a predator on its prey will be negative, and an increase of the prey will have a positive effect on the predator).

This format allows investigation of the role of nonadaptive selection in forming and maintaining community structure, without specifying the absolute values of the *a*s. In most cases we are interested primarily in the signs, and also the relative magnitudes, of the elements in the matrix. Then, the QSS procedure described in the text (chapter 2) and in section 10.12.2 allows us to evaluate the relative stability of matrices of different structure (signs and relative interaction values) measured by the proportion of the random matrices that are locally stable.

10.12.1 Numbering of Species Is Arbitrary

In general it should not matter whether we call a species "number 1" or "number 4." As long as the numbers assigned to the community's species are within the sequence, $1, \ldots, n$, the order in which we assign numbers to the species should not matter. As long as the interactions among the species of each interacting pair are correctly depicted, the community matrices will have the same stability properties (eigenvalues); what is true of one arbitrary arrangement of the data is true of all arbitrary arrangements of the data. Another way to say this is that the eigenvalues of the matrix remain the same under permutations of the rows and columns.

Thus, in the 4×4 matrix

$$\begin{bmatrix} a_{11} & a_{12} & a_{13} & a_{14} \\ a_{21} & a_{22} & a_{23} & a_{24} \\ a_{31} & a_{32} & a_{33} & a_{34} \\ a_{41} & a_{42} & a_{43} & a_{44} \end{bmatrix},$$

we can choose to depict the completely transitive, hierarchical, fully omnivorous trophic chain by numbering the species in order from 1 (the highest-level consumer) to 4 (the lowest species, which is only consumed), as we have done in section 7.3; this numbering $(1, 2, 3, 4)$ gives us

$$\begin{bmatrix} - & + & + & + \\ - & - & + & + \\ - & - & - & + \\ - & - & - & - \end{bmatrix}.$$

As usual, the + and − entries indicate arbitrary values of the corresponding sign. The hierarchical structure is clear, approaching a triangular matrix if the positive and negative values are asymmetric.

But had we, instead, numbered the species in the trophic chain differently, starting at the top and going down, as $(3, 2, 4, 1)$, we would get the matrix

$$\begin{bmatrix} - & - & - & - \\ + & - & - & + \\ + & + & - & + \\ + & - & - & - \end{bmatrix},$$

or if we numbered the species as $(2, 4, 1, 3)$, we would get

$$\begin{bmatrix} - & - & + & - \\ + & - & + & + \\ - & - & - & + \\ + & - & + & - \end{bmatrix}.$$

In both of these latter two the meaningful biological structure is not so obvious, but in fact all three are identical networks and exhibit exactly the same QSS.

10.12.2 Calculating QSS for a Matrix

In section 2.2 we discussed the rationale for QSS and the way that it is calculated (Allesina and Pascual 2008; Borrelli 2015). QSS is the percentage of randomly generated matrices with a given sign structure that are locally stable. This is a well-defined number, but its value needs to be interpreted with caution. We have alluded in the text to the fact that QSS does not tell us what proportion of these stable matrices are feasible (i.e., where all species have positive population sizes at equilibrium). We assume that where there are many stable possibilities, there will be a corresponding greater number of feasible matrices. In addition, the value of the QSS depends on the range of random numbers we supply for the randomly generated interactions. Obviously if we supplied only tiny numbers relative to the diagonal values, almost all matrices would be stable; species interactions would have negligible effects on population dynamics. Relatively huge values, on the other hand, would destabilize any matrix that was not strictly qualitatively stable (cf. Barabás, Michalska-Smith, and Allesina 2017). Clearly this is not the world we live in, and a range of more moderate numbers is used to obtain informative results.

For example, in Borrelli's (2015) simulations, the values for predator and prey are asymmetrical and drawn from different uniform distributions: the prey-on-predator value ranges over 0–10, whereas the predator-on-prey value is a random number from –1 to 0, as are the intraspecific values on the diagonal. The "simple matrices" from Borrelli and Ginzburg (2014) — see examples in section 7.3 — use the same numerical ranges, except that the diagonal values are all a constant value of –1. In both of these studies alternative combinations of values of the ranges also gave broadly similar qualitative results.

In general, using a variety of different value ranges yields qualitatively similar *relative* QSS values among the matrices, but the absolute values of QSS are less meaningful. Thus, in investigating the effects of asymmetry and different network topologies on QSS, we are able to make reliable qualitative generalizations but not quantitative ones.

In this book we are particularly interested in the qualitative effect of asymmetry in interaction values on QSS, so we want to make use of high levels of asymmetry. Unless stated otherwise, in this appendix the default for *asymmetric* values in example calculations is that weak interactions are drawn from a uniform random distribution ranging from 0 to 0.01 and strong interactions from 1 to 5 (with the diagonal values ranging from –1 to 0). For *symmetric* interactions, the default is that all off-diagonal interactions are taken from the distribution of "strong" interactions.

10.12.3 A Common Perspective on Different Types of Interaction Matrices

In this book we ask a very simple question about community matrices. We want to know whether the structural features seen most commonly across all of these networks in nature plausibly reflect (at least to some degree) the action of nonadaptive selection on local stability.

Traditional theoretical investigations divide community matrices into various types, depending on what biological processes they represent. Different questions emerge as being of interest in the study of different kinds of community interaction systems. Ecologists have used a variety of theoretical approaches, graphical representations, and terminology to describe consumer-resource (food web) networks, competition networks, and "bipartite" (e.g., mutualistic) networks (Mittelbach and McGill 2019). Researchers have tended to work on only one kind of system at a time, and the literature is somewhat fragmented. Gary Mittelbach and Brian McGill (2019) followed tradition in initially treating in separate chapters the theoretical approaches to the three main types of network just listed. They acknowledged that a full view of community interactions would unite all of these perspectives but admitted that this is currently a challenge. However, we are asking a much more specific question, about local stability, which allows us to treat the primary types of networks in the same way.

PREDATOR-PREY (FOOD WEBS). In section 7.3, we depicted a fully omnivorous trophic chain; this is a good place to start. Such a chain can be depicted as a fully transitive hierarchy in an approximately triangular matrix:

$$\begin{bmatrix} - & + & + & + \\ - & - & + & + \\ - & - & - & + \\ - & - & - & - \end{bmatrix}.$$

This has a QSS of 0.68 under this appendix's default asymmetry of interaction values, and 0.44 when the values are symmetrical. (As usual, it would have a QSS of 1.00 and be stable for all values above the diagonal if the values below the diagonal are all 0. But in this case, we have given them random nonzero values.)

We also know that a pure trophic chain looks like

$$\begin{bmatrix} - & + & 0 & 0 \\ - & - & + & 0 \\ 0 & - & - & + \\ 0 & 0 & - & - \end{bmatrix}.$$

This matrix is qualitatively stable (QSS = 1.00 in all cases). The two matrices look very similar, except for the presence of many zero values in the qualitatively stable chain.

Now, consider another variant of the omnivorous chain, in which the predators feed most strongly on the immediately lower species and less strongly on the others as they become more distant (see figure 10.4).

That is, the strength of the interactions decreases in a regular way down the chain. For example, predator 1 has strong interactions with species 2 (solid arrow), less strong interactions with species 3 (dotted arrow), and no direct interaction with species 4. If we denote the interactions as strong (S, s) and weak (W, w), with the lowercase letters indicating the relatively weaker interactions represented by the dotted lines, this network becomes

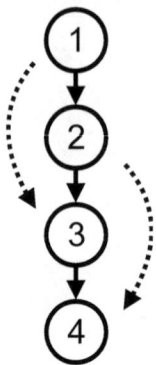

FIGURE 10.4. Omnivorous trophic chain in which intensity of predation attenuates with trophic distance

$$\begin{bmatrix} - & S & s & 0 \\ -W & - & S & s \\ -w & -W & - & S \\ 0 & -w & -W & - \end{bmatrix}.$$

The lowercase ss and ws occupy spaces that hold 0s in the pure chain. That is, this matrix now appears intermediate in structure between the transitive hierarchy and the pure chain; if the lowercase values here are 20% lower than the uppercase values, the QSS is 0.98 with asymmetry and 0.73 in symmetry. These QSS values are intermediate between those of a pure chain (1.00, 1.00) and a fully transitive hierarchy (0.66, 0.44), as expected. In general, predator/prey chains appear as species-interaction pairs of opposite sign, varying across highly stable pure chains and moderately stable transitive hierarchies (the latter depending on the degree of asymmetry in interaction values). As long as no lower species also feeds "upward" (meaning that transitivity is broken in such a way that positive signs appear on both sides of the diagonal), trophic chains should resemble the triangular matrix in structure and tend to be reasonably stable under strong asymmetry.

COMPETITION COMMUNITIES. The way that competition networks are typically represented in the literature tends to obscure the fact that they are structurally very similar to the food web matrices just discussed. A popular depiction of interspecific competitive interactions is based on the relative competitive ability of species, measured by the outcome of pairwise experiments among them (Keddy and Shipley 1989; Keddy 2001; Kinlock 2019). The results can be used to construct a binary interaction matrix (1s and 0s to indicate the winners and losers, respectively, in competitive contests). It is easy to see that such a matrix is equivalent to a sign matrix, as we illustrated for food chains, and is sometimes presented as such (Keddy 2001). As usual, the values of these positive or negative interaction coefficients need not be symmetrical. (Note: Throughout, we use the term *asymmetry* to refer to the relative *values* of the interaction coefficients for a species pair. In the competition literature, paired interactions between two species that are of *opposite signs* are often called "asymmetric"; what we call asymmetry of the *values* of interaction coefficients is sometimes called "imbalance" (Kinlock 2019). The two kinds of asymmetry are not always clearly distinguished in discussion of competition matrices, because the evidence suggests that opposite signs across the diagonal and value asymmetry are positively associated in competition matrices (Keddy 2001, 206–8).

Shipley's (1993, fig. 3c) example of a four-species "completely transitive" competition matrix, when expressed as a sign matrix, is exactly the same as our fully omnivorous trophic chain, discussed earlier. Both cases form a fully transitive hierarchy. The biological processes underlying the sign structure and asymmetry are different, but the form of the matrix is the same for both, and its QSS can be evaluated the same way. In competition matrices, transitive hierarchies with asymmetry of interaction values will resemble triangular matrices and be somewhat QSS stable. Some kinds of deviations from transitivity, however, will result in same-sign pairing across the diagonal and tend to lower QSS.

BIPARTITE NETWORKS. A bipartite network consists of two types of species that interact directly only with members of the opposite type. Plant-pollinator and plant-disperser mutualistic networks are well-known examples. Extensive study of mutualistic network architecture has revealed that in nature they are characterized by asymmetry and "nestedness." (A third frequently observed property, "modularity," is discussed in section 7.4.5.) Here, we show that nestedness and asymmetry in these networks are actually versions of what we have been calling triangular or hierarchical matrix structures with asymmetric values.

A bipartite network is usually depicted as a diagram similar to figure 10.5.

Species of, say, pollinators are across the top, and the plant species with which they interact are across the bottom. Each line depicts the pairwise interaction between two species. For example, animal species 1 pollinates plant species 5, 6, 7, and 8. (There do not have to be equal numbers of plants and animals, but the example is clearer if there are.) We can easily turn this into an 8 × 8 interaction matrix of the kind we have been discussing, but before we do, let us see how in the literature one demonstrates that this is a

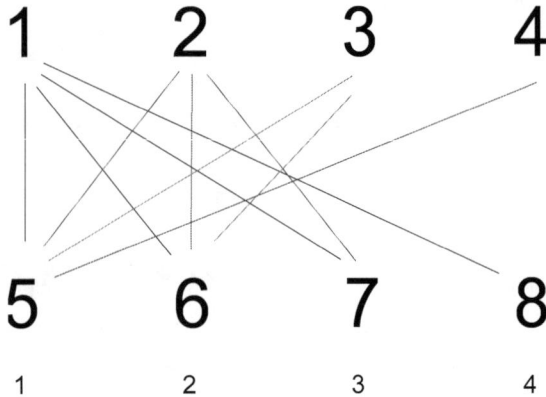

FIGURE 10.5. Bipartite network

nested hierarchy. The convention is that the animals are numbered from the most (1) to the least (4) generalized, the plants are numbered the same way (the small numbers in the diagram), and they are plotted against one another, with squares or other symbols to show the interactions (Bascompte et al. 2003). In our example, pollinator species 1 pollinates all plant species, and plant species 1 is pollinated by all animal species, and so on, creating a nested hierarchy. Figure 10.6 is an example of perfect nestedness.

If we use the initial diagram to make an interaction matrix for these eight species, and assuming that all interactions are positive (mutualistic) for both participants, we get

$$\begin{bmatrix} - & 0 & 0 & 0 & + & + & + & + \\ 0 & - & 0 & 0 & + & + & + & 0 \\ 0 & 0 & - & 0 & + & + & 0 & 0 \\ 0 & 0 & 0 & - & + & 0 & 0 & 0 \\ + & + & + & + & - & 0 & 0 & 0 \\ + & + & + & 0 & 0 & - & 0 & 0 \\ + & + & 0 & 0 & 0 & 0 & - & 0 \\ + & 0 & 0 & 0 & 0 & 0 & 0 & - \end{bmatrix}$$

Since we can use an alternate numbering for the species in the diagram (changing 1, 2, 3, 4, 5, 6, 7, 8) to (1, 2, 3, 4, 8, 7, 6, 5), this matrix is equivalent to

$$\begin{bmatrix} - & 0 & 0 & 0 & + & + & + & + \\ 0 & - & 0 & 0 & 0 & + & + & + \\ 0 & 0 & - & 0 & 0 & 0 & + & + \\ 0 & 0 & 0 & - & 0 & 0 & 0 & + \\ + & 0 & 0 & 0 & - & 0 & 0 & 0 \\ + & + & 0 & 0 & 0 & - & 0 & 0 \\ + & + & + & 0 & 0 & 0 & - & 0 \\ + & + & + & + & 0 & 0 & 0 & - \end{bmatrix}$$

This is our familiar interaction matrix, with many extra 0 entries because members (e.g., pollinators and plants) do not directly interact with other members of their type. From this we can intuit that when interactions are very strong (compared to interactions of species with themselves, on the diagonal) the matrix will be very unstable ($QSS \approx 0$), but if all the interactions are weak it may have reasonably high stability ($QSS \geq 0.50$). The matrix will also be relatively stable if interactions are strongly asymmetric ($QSS \approx 0.40$). So weak values in general and asymmetric values where the corresponding interaction is strong are the predictions for what would be favored by nonadaptive selection. This specific pattern has been observed to be prevalent in natural mutualistic communities (Jordano 1987; Bascompte, Jordano, and Olesen 2006).

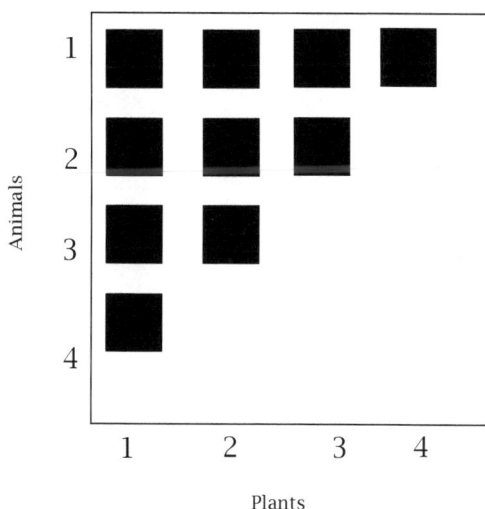

FIGURE 10.6. Depiction of interaction network with perfect nestedness

Note that in this analysis it is mostly the strength of interactions and the degree of asymmetry that have an effect on QSS. Why, then, is some degree of nestedness so often seen? Some deviations from nestedness increase stability, others lower stability, so nestedness per se is not obviously the key to stability in these matrices. Nestedness may arise from processes that have nothing to do with nestedness conferring stability, and nestedness may even be generally detrimental to local stability (Staniczenko, Kopp, and Allesina 2013; Suweis et al. 2013). Nested communities have a few specialists and many generalists. Our suggestion, from the perspective of nonadaptive selection, is that (usually imperfect) nestedness frequently appears or is preserved under asymmetry because of the difficulty of maintaining multiple generalists. Adding a generalist requires adding a large number of interactions, but adding a specialist adds mostly 0 entries. This is consistent with the observation that community *connectedness* (the proportion of realized interactions out of the total number possible) decreases with species number (Suweis et al. 2013). Nonadaptive selection against communities with too high a proportion of generalists is all that would be required to maintain the prevalence of nestedness in mutualistic communities.

SUMMARY. The major types of networks of community ecology can all be represented as matrices of the type that we analyze for local stability throughout this book, involving pairwise interactions among the species. This allows us to seek structural patterns that are common to these different types of communities. From the perspective of local stability, features that are widely considered to be characteristic of all of these types of communities in nature roughly correspond to the quasi-triangular, hierarchical, and asymmetric matrix structure that would be favored by nonadaptive selection on local stability.

10.13 Species Selection: How the Concept of Species Selection Has Developed over the Last 50 Years, and Why It Still Doesn't Quite Work

In this appendix we develop the perspective that nonadaptive selection can, under certain conditions, lead to a strong link between ecological and species-level characteristics, thereby providing a mechanism for species selection. As we have remarked (see section 3.2), selective processes involving entire species, in which "fitness" is based on their propensity to speciate or to suffer extinction, were not thought to be controversial by Darwin and the early synthesis, but they were also not thought to be of much importance (Fisher 1930; Gould 2002). In the mid-1970s species selection reemerged in the growing macroevolution literature (Stanley 1975). Species selection has been seriously discussed for almost 50 years, yet we still do not have an accepted theory that produces causal, mechanistic (nonhistorical) explanations for clade patterns. Opinions and concepts have varied considerably among researchers. Almost everyone agrees that at least sometimes, something like species selection really happens, and that there are macroevolutionary patterns that cannot be explained by simple extrapolation of within-species evolutionary trajectories (Grantham 1995; Jablonski 2008, 2017).

In this appendix we first describe the prospect for selection per se at the species level. Then we follow with a brief history of ideas about how such species selection should be interpreted. Finally, we introduce our solution to the major difficulties we identify and briefly discuss the use of multilevel selection theory in the context of species selection.

10.13.1 Selection at the Species Level Is Plausible, but Not the Whole Story

Throughout this book we have considered the Price equation (appendix 10.4) as a general description of any selection process; recall that the Price equation can apply to models of adaptive natural selection as well as to those of nonadaptive selection, in both living and nonliving populations or systems. From this perspective there isn't any reason to doubt that a process of selection can possibly occur among species. However, this does not mean that there is no difference between this species-level selection process and those that we have discussed in this book up to now, involving ecological entities. We argued in chapter 3 that species and clades are not ecological entities but are historical constructs based on genealogy. The context for historical events that involve such genealogical entities is not a geographical location but rather a particular genealogical scheme to which they belong.

Species are seen, on occasion during their historical existence in a genealogy, to give rise to new species ("reproduce") and to go extinct ("die"). A species thus has a historical beginning and a real or potential end, and the frequencies of different kinds of species within clades (which are global constructs of genealogically related species) could change over time as a result of selection at the species level. If a species' probability to produce a daughter species is S and the species' probability of extinction is E, and these probabilities are associated with any characters exhibited at the species level, then we have a basis for assigning relative "fitness" values for use in the Price equation (appendix 10.4). To be clear, selection in the Price equation does not require "emergent" species properties; any properties that the species exhibit at their level are allowed (Lloyd and Gould 1993).

With regard to the Price equation, it is possible to have selection occurring among a group of entities of all kinds, living or nonliving, and in "populations" that are defined in various ways, not necessarily as ecological populations. Thus, *selection* is extremely general. But it is only *meaningful* if the fitness values at the focal level have meaning within the context of the "population." So, for example, one could make a population comprising three mice of different species and on different continents, a duck, a wooden fence, an ice cube on Mars, and a bacterium in a deep-sea vent. One could assign them relative fitnesses and thus enable selection to work in this "population." But it is unlikely that there would be any rational basis to arrive at those relative fitnesses, since there is no meaningful way to compare the various population members. So, yes, technically selection could be occurring, but it would mean nothing. Likewise, selection among species can clearly be observed, but only if the assignment of fitnesses makes some kind of general sense for species in a particular clade does that selection mean anything other than a description of a series of historical events.

Further, a species-level trait will not evolve under selection unless it exhibits heritable variation in fitness (Lewontin 1970; Jablonski 1987). Note that mere heritability of the trait itself (level of resemblance between parent and offspring) is not sufficient. The trait's phenotype must be *reliably* associated with differences in relative fitness. A trait may be highly heritable, but if its phenotype is not associated with fitness, it will not evolve. On the other hand, a trait with low heritability that is nevertheless reliably associated with fitness can evolve by selection, just not as efficiently as it would if heritability were higher (chapter 3). Thus, the reliability of fitness/phenotype associations at any and all levels to predict species-level differences in relative fitness emerges as a critical feature of the species selection process. We will see that this has been a major difficulty for different views of species selection.

To sum up, there is certainly a potential selection process among species of a clade; however, although we can often observe a kind of selection process, it does not follow that there ever is adaptive selection at or above the species level, nor does it by that token necessarily provide any general explanation for repeated patterns of clade change. As we argue throughout the book, a selection analysis by itself does not constitute a causal analysis of the source of fitness differences (chapter 3).

10.13.2 Species Selection, Phase 1: Description of History

At the risk of oversimplifying, we can discern in the conceptual development of species selection three main temporal stages. The "original" species selection in the modern era was that of Steven Stanley (1975, 1979).
Stanley concentrated on the species-level characters involving differential
S and E, that is, the observed differences in species-level fitness that constitute the apparent *selection* at the species level. He regarded these species-
level fitness components as being established and fixed for a species at its
origin, independently of any trends ruled by selection at any lower level.
In Stanley's view, the only viable goal in studying species selection was to
provide a historical explanation for observed instances, rather than to advance a *mechanistic* explanatory scheme based on causes independent of
space, time, and incidental historical circumstances. Stanley did not claim
to demonstrate a mechanism for producing species-level adaptations and
regarded natural selection (at all levels) as unable to provide anything
other than a description of history (Stanley 1979, 192–93). Alternative
formulations in this tradition came to similar conclusions (e.g., Cracraft
1982). Species selection was selection, but it wasn't actually Darwinian.

10.13.3 Species Selection, Phase 2: Adaptive Evolution at the Species Level

The second phase arose in reaction to this view and was based on a number of related claims. Species selection was reconceived as a "hierarchical
expansion" of the *full* Darwinian theory of adaptation by natural selection.
Species played the part of organisms in Darwinian theory, clades the part of
ecological populations, and so forth. "True" species selection was regarded
as adaptive and as forming "adaptations" at the species level (Vrba 1980,
1984; Vrba and Eldredge 1984; Vrba and Gould 1986; Gould 2002).

Unfortunately, some critical issues in this hierarchical expansion were
glossed over and never successfully resolved. Most significant is the fact that
clades make a poor, if not impossible, analog of a shared population (see
Damuth 1985; Grantham 1995). Clade membership is based on genealogy,
not the occupancy of a localized, shared environment. Evolution by selection
at any level requires heritable variation in fitness associated with at least one
target character. As we argued in chapter 3, *adaptive* evolution also requires
the population to occupy a shared, local environment. Such an environment
cements the association between character states and local relative fitness
and is the mechanism of adaptation to *local* conditions. Without a shared

environment, there is no sure relationship between a character state (at any level) and fitness at the species level. Where the "population," such as a clade, does not occupy a shared local environment for the units under selection, such selection cannot be adaptive. (As discussed in chapter 3, *nonadaptive* selection is immune from this requirement for a shared environment, because adaptation is not a component of the nonadaptive process.)

Much of the conceptual development of species selection during the second phase sought to erect classification schemes involving kinds of characters at different levels (e.g., "individual," "emergent," "aggregate"), certain causal relationships among them ("upward," "downward," "effect"), and various "kinds" of fitness (e.g., "reducible," "emergent," "nonreducible"), that in some combinations either certainly would, or in other combinations could not possibly, support selection leading to adaptive species selection. However, the conversation leaned heavily on views of natural selection common among neontologists working at the organismic level, where shared environments are an (often unstated) assumption. The resulting schemes were interesting exercises but did not quite do the job of showing how adaptive evolution at the species level was possible, let alone of paramount importance. This was because no general way was found to resolve the issue of the lack of a shared environment and the resulting difficulty of associating organisms' or species' character states with heritable variation in species-level fitness.

A further limitation was the lack of a generally agreed-upon definition for an *adaptation* at or above the species level. Such an adaptation would presumably require a demonstration that the evolution of a character state or novel trait at the species level results in an improved "fit" between a taxon and its environment (which it doesn't really have). In fact, investigation of the existence of such adaptations was largely ignored. In contrast to Darwinian natural selection, which at a stroke explained a vast array of self-evident adaptive fits between organisms and their environments, there was no a priori catalog of species-level adaptations, or adaptive fits between species and their clades, demanding a mechanistic explanation. Instead, most people have been interested in species selection to answer questions about the patterns of change caused by the selection, not questions of adaptation per se.

Our explanation for the failure of phase 2 concepts to provide a useful theory of adaptive species selection is, simply, that we think it extremely unlikely that adaptive selection can occur under these circumstances. This is due not only to the fact that species and clades are not ecological units, but rather are historical constructs based on genealogy. It is also, specifically, that there is no ecological population to which the species belong, and the species of a clade ordinarily do not share a local environment. Finally, the lack of a well-accepted definition of what a species adaptation might be is, we think, significant. In keeping with our practice in the rest of this volume, we do not claim that there can *never* be a widely accepted definition of species-level adaptations, but we think that even if such a concept were to become established, it would necessarily be highly restricted and of little general significance. So phase 2 *adaptive* species selection is, by definition, Darwinian, but apparently never, or almost never, occurs.

10.13.4 Species Selection, Phase 3: Species Selection Is "Reinvented"

More recently there has been a move to "reinvent" species selection, which in a sense returns species selection to its roots (Rabosky and McCune 2010). The "new" species selection takes as its goal the study of the statistical association between organismic traits and diversification rates of species within and among clades, using detailed molecular phylogenies of (mostly) extant species. Such studies regard selection among species (based on their species-level traits) as relatively unproblematic—"species selection" (or "species sorting") is the observed selection; there is really only one kind.

The prospect of general explanations in phase 3 is based on the expectation that characteristics of the organisms belonging to each species *reliably* generate, or at least reliably correlate with, differences in species origin and/or extinction rates. Causal influences must percolate up from the dynamics of the ecological entities within the species to the level of all the species of the clade (or, at least, all the species that are considered subject to the supposed selection process). Moreover, because the clade is a historical construct and has no necessary ecological characteristics, lower-level ecological units exerting reliable effects on speciation and extinction rates must do so consistently in the face of ecological, geographical, and temporal variation in environments and circumstances within each species and across the clade. Such ecological and temporal variation can break the causal chain from the supposed source of differences in observed fitnesses (effects among ecological subcomponents) to the level at which selection is acting (among species).

As an analog of—or extension of—adaptive natural selection, the situation is worse than simply not having a causal connection between characters or fitnesses at multiple levels. Even if there is a stable, cross-species causal relationship between an internal, ecological character and diversification rates in the clade, interpretation is still uncertain. For example, imagine that a species' probability of extinction depends primarily on the average probability of extinction of its constituent ecological populations. The species is composed of many such populations, possibly spread over a wide geographic area and extending sometimes for significant periods of time. Therefore, it is likely that these populations will experience a range of different biotic and abiotic environmental influences. If a trait value of individuals contributes to population stability in some populations, but the same trait value confers instability in other populations, then whether the trait contributes to high or low extinction at the species level depends on the historically contingent deployment of the species' populations in space and time. Furthermore, the species-level trait value depends on whether there is any tendency for the trait to evolve adaptively to local conditions at lower levels throughout the species' lifetime.

Niles Eldredge et al. (2005) argued that the ubiquity of such variation in the relationship of traits to fitness across different environments within a species is the primary source of evolutionary stasis. Species are constantly being pulled in different evolutionary directions because of their being deployed over multiple environments, in which different traits are favored. Thus there is ordinarily no net adaptive change, and variation across the species' range is neither constant nor predictable. Thus, "stasis" represents not the absence of evolution but the effect of inconstancy of the relationship between trait values and fitness. Accepting this argument about stasis would underscore the difficulty of judging the reliability of organismic traits in predicting *species-level* selective properties from incomplete historical and ecological data about a given species.

Although there are now many examples of statistical associations between organismic traits and diversification rates (Stanley 1979; Rabosky and McCune 2010; Jablonski 2017), this bounty comes at an explanatory cost. In phase 3 species selection, there is still no shared environment for the species that will cement their organismic character(s) consistently to a specific species-level character or the species' fitness at its level. This means that, similar to phase 1, phase 3 species selection can describe a particular, historical diversification event, but that observation in itself is not evidence of any *general* pattern applying to all species that exhibit the target character states. It is possible that an observed statistical association of extinction or diversification with an organismic character alone will suggest a plausible causal model of species fitness variation. But it is unlikely that we will have the detailed information (especially for fossil species) that would be required to verify the general reliability of the trait's association with fitness or to infer the domain over which such a model is expected to be valid.

Thus we see that the same issue that bedevils the concept of adaptive species selection—lack of a shared environment—limits the explanatory power of third-phase species selection as well. In both cases the distributions of traits and adaptations among the individuals and populations of a species do not automatically reliably generate fitness values or fitness-related characters at the species level.

It is perhaps no surprise, then, that the interpretation of phase 3 species selection is challenging, even should phylogenies be sufficiently robust and statistical techniques adequate (Rabosky 2010; Rabosky and Goldberg 2015). Many ecological traits are correlated or conflict with each other, and they may evolve within different species at different rates (Rabosky and McCune 2010). The same traits might have different effects in different clades (Upham, Esselstyn, and Jetz 2020).

Phase 3's effective restriction to the historical mode of explanation has led to an expanded view of the term species selection (or sorting) as one part of a descriptive, historical body of concepts, including other historical sorting processes, such as widespread directional environmental change, episodes of mass extinction, and effects of interactions with distantly related clades (Eldredge 2003; Jablonski 2017). In phase 3, we are led back again to the vast complexity of history.

10.13.5 A Nonadaptive View of Species Selection and Its Components

From the beginning of the modern discussion of species selection, a few cases have been widely regarded as good candidates for this process, under most definitions. One of the most widely known and well-studied is the effect of the larval ecology of marine mollusk species on speciation and extinction rates. Benthic marine invertebrate species exhibit one of two primary kinds of larval development: free-swimming larvae that spend a long time living and feeding in the plankton before settling, and which thus can disperse widely (*planktotrophic*), versus *nonplanktotrophic* larvae that spend little or no time in the plankton and have limited dispersal ability (Jablonski and Lutz 1983; Jablonski 1986). This difference in dispersal ability affects observed differences in speciation and extinction probabilities among fossil mollusk species. Current thought sees this effect operating primarily through the influence of planktotrophic larvae on the maintenance of large geographic ranges, which contribute to lower extinction rates (Jablonski and Hunt 2006). (Although the rate differences based on ranges are measurably heritable at face value, this must be due to something that the species possesses internally. The mechanism based on the type of larvae of a species is one example of such an internal mechanism.) In this case the larval type of the species determines a difference in higher-level fitness, through a straightforward ecological mechanism. In other words, an a priori mechanism involving an ecological-level trait (larval ecology) yields an effect on diversification or extinction rates that is largely insensitive to variation in the environments experienced and temporal occurrence of the ecological units belonging to the taxon. This character state is thus reliable, in the sense that we have been discussing. It has a kind of universality, with respect to an identifiable domain, as do nonadaptive lawlike generalizations that we have been discussing throughout this book.

Other strong candidates for mechanistic species selection share the same features in being based on characters or causal schemes whose effects at the species level are nonadaptive, "universal," and independent of local or temporal circumstances. Such candidates include (1) the wide occurrence and maintenance of sex, which presumably operates through long-term advantage to most species, regardless of circumstances (Fontaneto et al. 2012), and (2) the presence of sexual selection within ecological populations of the species, which is thought to increase the probability of population differentiation and rates of speciation (Lande 1981; Barraclough, Vogler, and Harvey 1998). In such cases, the generality of the result comes from a factor whose relationship with species-level fitness is consistent across local environmental or temporal variation within a species.

The target character may or may not be fixed within a species; its ability to generate consistent species-level fitnesses comes not from any lack of variation in the character state, but rather from a consistent relationship between the character and species-level fitness differences that does not vary in space or time. There is thus a strong nonadaptive component to the maintenance of that relationship. The effect of the character on species-level fitness is a lawlike generalization, either a domain-specific necessity or the result of a nonadaptive selection process—whether explicitly recognized as one or not.

This leads directly to our suggested solution in chapter 8 for many of the difficulties encountered in species selection research. Nonadaptive selection processes involving the ecological entities within species can, in special circumstances, lead to a secure connection between ecological properties and species-level properties, allowing for a mechanistic species selection. This is discussed more fully, with an example, in chapter 8.

10.13.6 Multilevel Selection Theory and Species Selection

Some have postulated that multilevel selection (MLS) theory (Heisler and Damuth 1987; Damuth and Heisler 1988), as used to perform a selection analysis (see chapter 3), can be a general basis for species selection theory (Arnold and Fristrup 1982; Jablonski 2008, 2017). However, MLS may not be as useful as it may at first seem. Versions of MLS approaches, as derivations from the Price equation, address only the mechanics of selection. Thus, by itself an MLS analysis cannot solve any issues of causal relationships between levels—though it may well suggest possibilities to investigate further.

MLS derives from the Price equation (appendix 10.4). In the full Price equation the selection (covariance) term contains the fitness of z, a character borne by the focal entity and of no other entity. Everything involving the selection term must be in terms of the fitness (w) of states of z. Higher levels than that of the focal level that bears z appear only as effects on z's *fitness* (w). Going the other direction, lower-level selection appears only in the expectation term, but this time, only in terms of an effect on the *value* of z. So selection is always "about" only one level, the one to which z belongs, as modified by (nonsymmetrical) influences from other levels. That is, the higher level modifies the fitness of z, and the lower level modifies the value of z. Neither of the forms of MLS is capable of directly representing simultaneously separate selection processes at more than one level.

If individuals are the focal level, one can detect an effect of "species membership" on individual fitness, but one cannot conclude that such an effect necessarily derives from selection (differential speciation and/or extinction) at the species level (cf. Lloyd and Gould 1993). If, on the other hand, species are the focal level, lower-level selection among individuals can be represented (in the expectation term, or expansions of it) as an average change in the species-level character (z), but this does not appear as a process of selection, only its effect (Damuth and Heisler 1988).

However, MLS is useful in a special situation in which both the individual and the species character are the "same" (e.g., mean body size) or can plausibly and necessarily be related to each other. Then the Price equation permits a quantitative partitioning of overall clade trends in individual characters into components due to within-organismic lineage evolution and species-level change (Simpson 2010, 2013; Rankin et al. 2018). If, as we suggest in section 8.2.2, species selection is best viewed in a context of nonadaptive processes, individual and species-level characters and their effects on fitness may frequently be "the same." Thus MLS approaches may take on increasing value and can help constrain subsequent causal interpretations. Note, however, that techniques other than MLS can also accomplish the same partitioning (Jablonski 2017).

10.14 What Nonadaptive Selection Is Not

The landscape of evolutionary and ecological theory includes many processes in addition to adaptive natural selection. We are not positing any
novel or exotic kinds of processes intrinsic to or internal to organisms,
such as vitalistic mechanisms, saltation (genetic revolutions), or orthoge-
netic forcing. The inherent properties that multispecies biological systems
exhibit are intrinsic properties or logically deduced consequences of their
ordinary biological properties and interactions, and the mechanism of
change is simply selection (per se) as it has been generally understood.
The only thing that is distinct about nonadaptive selection is that due to
the nature of the properties conferring "fitness" or stability, this selection
does not result in, and is not driven by, local adaptive evolution.

Nonadaptive selection may nevertheless bring to mind a variety of explanatory schemes that either are well established in evolutionary biology
or have been previously advanced for understanding evolution at the macroscale. Here, in no particular order, we briefly differentiate our concept of
nonadaptive selection from other processes most likely to cause confusion.

10.14.1 Randomness and Neutrality: Nonadaptive Evolution,
but It's Not Selection

The term *nonadaptive* is frequently applied to processes that cause evolution but are *not selective*. These include a variety of processes that are characterized as stochastic or ones that are neutral. Examples include mutation,
recombination, genetic drift, the evolution of neutral alleles, and similar
kinds of processes (Van Valen 1960; Darwin, as cited by Dykhuizen and
Hartl 1980; Lynch and Hill 1986; Gittenberger 1991; Hubbell 2001; Hubbell and Lake 2004; Lynch 2007a, 2007b). In contrast, nonadaptive selection
(like all selective processes) acts through its effects on the fitness (suitably
defined) of units undergoing selection. Without a selective process evolutionary change can still occur, but this is not the subject of our essay.

10.14.2 Selection of Correlated Traits

Traits may be correlated through pleiotropy or genetic linkage. In such
a case evolutionary change in a trait that is the actual target of selection
may cause changes in a correlated trait that otherwise is not the target

of any selection process (Lande and Arnold 1983). So the evolutionary change of the correlated trait is not adaptive. But this change is the result of *adaptive* selection on the other, target trait(s). In our sense there is no separate *nonadaptive* selection process going on, involving the fitness of the correlated trait, only the adaptive process involving the other trait. The correlated response itself is, like the change in the target character, specific to the nature of local conditions that underlie the adaptive process. Nonadaptive selection in our sense is not occurring.

10.14.3 Nonadaptive Is Not (Necessarily) Maladaptive

In some cases people use the term *nonadaptive* to mean *maladaptive*. Selection optimizes relative fitness but is not guaranteed to optimize the absolute function or value of any specific character (Brady et al. 2019). *Maladaptation* suggests that the realized level of "adaptedness" is lower than the optimum that would be expected in a *specific local environment*, based on an assessment of ideal alternatives. Maladaptation can occur for many reasons, including genetic architecture, functional trade-offs among traits, gene flow, environmental change, and even variation in adaptive coevolution across the landscape (Michod 1986; Thompson, Nuismer, and Gomulkiewicz 2002; Brady et al. 2019). But all of this has to do with impediments to, or different definitions of, adaptedness to *local conditions*. Nonadaptive selection in our sense does not recognize the environmental conditions that result in adaptive evolution to the local environment, nor does it select against entities that are maladapted (in some sense) to the local environment per se. The incidental effects that nonadaptive selection may have on "adaptedness" of any evolutionary or ecological unit can be positive, neutral, or negative, depending on the situation; nonadaptive selection itself is unaware of relative adaptedness to a specific environment.

10.14.4 Nonadaptive Selection Processes Are Not Some Kind of Higher-Level Adaptation

Note that we do not imply that the conditions for nonadaptive selection to work on a given entity have evolved by some kind of adaptive evolution for the purpose of facilitating or initiating the nonadaptive process. Evolution by nonadaptive selection is not itself an adaptation or a solution to any local adaptive challenge. Nonadaptive selection is agnostic with respect to what kinds of processes form the units upon or within which it may be operating.

10.14.5 Sexual Selection

Frequently, evolutionary biologists make a distinction between *natural se-lection* and *sexual selection*. Sexual selection is that due to differences in mating success, contrasted with other sources of adaptive selective forces (Arnold and Wade 1984), and can lead to unusual and complex evolutionary dynamics and spectacular phenotypic and behavioral adaptations (Lande 1980; Kirkpatrick 1982; Diamond 1986). The differentiation of natural and sexual selection originated with Darwin (1859), who considered that natural selection always acts to improve adaptedness to the local environment, whereas sexual selection may well lead to adaptations that seem to impair adaptedness (Mayr 1972). However, sexual selection is not a type of nonadaptive selection, in our sense. The local environment includes those factors that are relevant to both ordinary natural selection and sexual selection, and phenotypes interact with all of them. Endler (1986) described sexual selection and "nonsexual selection" as subclasses of Darwinian adaptive natural selection, and current models treat the effects of sexual selection as one component of fitness within an overall adaptive selection process (Arnold and Wade 1984; Sober 2000). Natural selection where sexual selection is present still results in optimizing relative fitness with respect to local conditions, including the prevailing mate choice preferences and sexual signaling characters. Thus, sexual selection is a type of adaptive evolution, but one that may favor adaptations that do not increase the fit to local, *nonsocial* environmental conditions.

10.14.6 Adaptive Parallelism and Convergence

Adaptation can of course cause parallel and convergent evolution (Agrawal 2017), but such evolution is obviously adaptive and thus contingent on—adaptation to similar environments (in the broadest sense) that either happens or not, depending on the existence of those environments and the local phenotypic variants that are present in them.

When we refer to the "universality" of a nonadaptive selection regime, we mean just that. Nonadaptive selection, taken by itself, operates the same way, and with the same relative strengths, throughout the domain within which it is applicable (see section 3.3 about biological "laws"). Adaptive evolution is also capable of producing similar results in different times and places. In such cases the similar characters involved have still evolved via historically contingent adaptive selective processes. This is evolution in response to similar environmental challenges that are (or have been) frequently encountered by different individuals, populations, and species. In some cases, the differential probability of producing different classes of mutations tends to result in evolution down preferential paths, but the selective process of favoring and preserving such novel variations is nevertheless adaptive. Homologous structures have the status of a shared genealogical history (Wagner 2014); analogous structures perform similar functions with different historical origins. But the origins of both kinds of characters are all unique historical events, in the context of local environments. Universality, in contrast, does not refer just to something that happens (or happened) often.

Adaptive evolution can limit diversity of organismic form because of such historical considerations. But adaptation also explains what diversity we see with respect to the diversity of local environments. Nonadaptive selection, on the other hand, cannot explain diversity, only widespread similarity.

10.14.7 Not Simply Selection on Stability

Selection on stability properties can be either adaptive or nonadaptive, depending on the origin of the effect of phenotype on stability. Although the phrase "stability selection" has been used loosely to refer to some of the explanatory schemes we are calling nonadaptive selection (Borrelli et al. 2015), we recognize nonadaptive selection only when the variation in stability arises from variation in inherent properties that do not differ across environmental contexts. Certainly entities can be subject to selection on their stability within a specific, historically unique environmental context. In such a case the stability property varies in its effects primarily as the result of an interaction between some phenotypic character and relevant characteristics of that context. Selection on adaptive components like this would lead to increased stability only in the specific local environment; in other environments this component of selection might favor different trait values. In contrast, nonadaptive components of selection on stability favor or penalize particular phenotypes in the same way regardless of environmental circumstances.

10.14.8 Internal Selection

Within evolutionary developmental biology there is a phenomenon called "internal selection" that may seem similar to what we are proposing but in fact is quite different (Whyte 1960; Arthur 1997). *Internal selection* is defined as occurring when individuals bearing genetic variants or novel mutant genes are retained or eliminated because of their genes' effect on internal developmental or functional coordination of the organism (Arthur 1997). As usually described, this is clearly an adaptive process; fitness comes from an interaction with a local environment and selects for novel variants that perform well under the local environmental conditions. The only distinction is that here, the environment with which the genotype interacts is the internal genomic environment and not (necessarily) the environment external to the organism. For adaptive selection it does not matter whether the environment is internal or external; it is still a local environment. Notwithstanding the value of making the internal/external distinction for studies of the evolution of developmental processes, internal selection is best considered a subset of Darwinian adaptive natural selection (Fusco 2001). In contrast to internal selection, nonadaptive selection does not cause the proliferation of novel features; in nonadaptive selection the fitness values depend on inherent properties that are themselves not subject to adaptive evolution. In effect, nonadaptive selection does not see *any* environment.

This is not to preclude the possibility that there may be nonadaptive selection processes operating within organisms during development. Should such processes exist, they would not have the properties generally now ascribed to internal selection, nor would they necessarily conflict with it.

10.14.9 Group, Multilevel, and Hierarchical Selection

The short answer is that group selection, as understood by the majority of biologists, is a kind of *adaptive* evolutionary theory, and the controversy surrounding the concept has overwhelmingly involved conflicting views of the effectiveness of different processes in determining the adaptations of individual organisms or in forming adaptations of higher groups as such (Lloyd 1994, 2008). We do not take a categorical position here on the circumstances under which *adaptation* can or does occur at levels higher (or lower) than the individual; our interest is in selective evolutionary mechanisms that do *not* lead to adaptation. So even though the nonadaptive selective processes that we discuss often involve entities other than individual organisms, those processes are not able to form adaptations at *any* level. The actions of nonadaptive processes are thus compatible with any kind of models of Darwinian adaptive evolution, including explicit multilevel selection or alternatives such as inclusive fitness approaches. If nonadaptive selection is operating in a given situation, it does so as a component of selection that does not vary with the environment. Furthermore, we focus our discussion on higher levels at which adaptive selection processes seem to be absent or weak. Of course it is possible that there is also adaptive selection going on at those levels; if so, discussion of it is beyond the scope of the present essay.

Before leaving the topic, it is worth noting that group selection has not always been controversial. Early considerations of selection among groups and species by Darwin, Fisher, and others took the existence of such selection—differential survivorship among entities higher or more inclusive than the organism—to be self-evident but of little consequence for *adaptive* evolution of organisms, except in special and unusual circumstances. This view has continued to prevail among most evolutionary biologists.

However, throughout a recent period of often acrimonious controversy, "group selection" has taken on an expanded meaning of a selective process that acts *on* or *through* or *because of* the existence of entities above the level of the individual organism, such that individual fitnesses (and the process of adaptation to local conditions) are thereby affected. Most generally, various kinds of population structure will lead to nonrandom interactions among individuals, which under the right circumstances can significantly affect the course of adaptive evolution of organismic characters (Wilson 1980; Arnold and Fristrup 1982; Heisler and Damuth 1987; Damuth and Heisler 1988; Goodnight, Schwartz, and Stevens 1992; Sober and Wilson 1998; Okasha 2006; Wade 2016). This expansion of the concept was originally driven by reaction to a specific kind of theory about population regulation promoted by Wynne-Edwards (1962, 1965, 1986), which attributed to group selection the evolution of specific *organismic* adaptations "for the good of the group."[11] It was the implications for the selective formation of individual adaptations (not Wynne-Edwards's now-questionable views on population regulation) that motivated and sustained the heated controversy (e.g., Maynard Smith 1964; Wiens 1966). The relevant point in the present context is that the issues in what came to be called the "units of selection" controversy were never about whether *selective processes* could involve supra-organismal entities, but to what degree those (and other group-dependent) processes could override selection among organisms and thus select for traits apparently disadvantageous to individual organisms (taken one at a time, e.g., "altruistic" behaviors). That is, they were primarily about the locus of adaptation (Lloyd 1994, 2008).

This expanded view of group selection brought the topic into apparent conflict with an influential alternative approach to explaining the formation of individual adaptive behaviors in structured populations, based on *inclusive fitness* (Hamilton 1964). The controversy left anything associated with selection above the organismic level under a cloud of suspicion (at best). However, it is now generally recognized that (as some had long suspected) inclusive fitness models and multilevel selection models are mathematically equivalent in terms of how they model selection mechanics (Hamilton 1975; Wade 1979, 1980, 1985; Gardner 2008; Marshall 2011; Goodnight 2013; but see Simon, Fletcher, and Doebeli 2012; Simon 2014). Much of the heat has now gone out of what was always a somewhat narrowly focused controversy. However, the key issue concerning the locus of adaptation is not solved by this unification (Wade et al. 2010; Wild et al. 2010; Jeler 2020). See appendix 10.13 for additional discussion of "hierarchical selection" in macroevolution, and appendix 10.6 for a fuller discussion of adaptation in the context of multilevel selection and inclusive fitness.

10.14.10 Major Evolutionary Transitions in Individuality

John Maynard Smith and Eörs Szathmáry (1995) described a series of seven major *transitions* in the history of life, at which higher-level individuals resulted from evolution among their formerly free-living components (see also Buss 1987). Well-known examples include the transition from prokaryotic to eukaryotic cells and the origin of multicellular organisms from free-living unicellular organisms. Interest in this topic has been high, for although major transitions of this nature are rare, some have been of enormous significance (Szathmáry 2015; Radzvilavicius and Blackstone 2018). As had many before them, Maynard Smith and Szathmáry recognized that in each case, loose associations or assemblages of separate individuals had evolved into fully functional "superorganisms" at a higher level, involving associated changes in information processing, transmission, and reproduction. Significantly, these transitions also involve the transfer of primary participation in adaptive evolution from the lower-level constituents to the new "superorganism" and the loss of ability among the lower-level units to adapt individually to local conditions experienced by the collective.

Thus, such evolutionary transitions are considered to be driven by adaptive evolution, and in part they consist of change in the locus of adaptation within the evolving system. As we emphasize throughout this book, study of selection dynamics alone is inadequate to identify and track the locus of adaptations. Study of major transitions raises many issues about defining adaptation and "tracking" adaptations (Griesemer 2000; Michod 2005; Okasha 2006; Clarke 2014; Shelton and Michod 2014a; Black, Bourrat, and Rainey 2020; Griesemer and Shavit 2023). However, *nonadaptive* selection processes, as described in this book, cannot directly drive such adaptive transitions.

10.14.11 "Gaia" and the Evolution of Biogeochemical Cycles

Earth exhibits large-scale negative feedback mechanisms that act to stabilize the planetary environment, and some of these involve activities of living organisms interacting with nonliving systems (Lenton 2016). For example, the composition of Earth's atmosphere is regulated by complex feedbacks involving both geochemistry and biology. James Lovelock and colleagues (Lovelock 1972; Lovelock and Margulis 1974; Lovelock 1979) drew attention to the importance of these biogeochemical links and cycles in maintaining a biosphere on Earth that is conducive to life. In a compelling metaphor, he described Earth's system as a giant organism he called *Gaia*.[12] This made it sound as if he were claiming that, from the time of life's origin, the biological component of such feedback loops somehow evolved (adapted?) *for the purpose* of continuing to support life as we know it. The Gaia hypothesis has never been about the possibility of adaptive natural selection at the biosphere level per se — that is, evolution among higher-level (or the highest-level) biological entities themselves. Rather, strictly speaking, it refers only to the origin, maintenance, and modification of geochemical feedback mechanisms. This topic is at first glance somewhat removed from our discussion in this book. However, the existence of global-scale adaptive processes is still an open issue.

At first, biologists lost no time in criticizing the Gaia hypothesis for its lack of a plausible evolutionary mechanism for such higher-level adaptive scenarios (Doolittle 1981; Williams 1992a; Dawkins 1999; Kirchner 2002). Defenders of Gaia responded that the Gaia hypothesis does not actually *require* that there be any higher-level adaptive selection (Margulis 1981; Hamilton and Lenton 1998; Lenton 1998; Wilkinson 1999; Ruse 2013). In the face of inorganic change, organisms may evolve new adaptations that cause them, in turn, to change the biotic and/or abiotic environment, leading to adaptive evolution of other organisms and/or modifying geochemical processes. Neither the origin nor subsequent elaboration of the biological component of such feedback loops is considered by most biologists to require anything other than ordinary organismic adaptive selection. Gaia has thus become a largely noncontroversial description of how life interacts with and modifies biogeochemical cycles, while requiring no organic evolutionary mechanisms other than ordinary adaptive natural selection, and most biologists have left it at that (Staley and Orians 2000, 52–53; Lenton 2016).

Nevertheless, a number of researchers, including some early critics, have explored anew the idea that some kind of higher adaptive processes could indeed be involved in forming Gaia-scale cycles and regulatory feedbacks (Doolittle and Inkpen 2018). Just because there is no *requirement* for higher-level adaptation to be involved does not mean that it didn't happen—if it were possible. Unfortunately, most such attempts have merely established that *selection* can happen at higher levels (which we support), but assume that adaptation automatically occurs as a result (with which we disagree). (See section 3.3 for a fuller discussion of higher-level selection versus higher-level adaptation.) Because the distinction between selection per se and adaptation is seldom drawn, it is often difficult to interpret proposed Gaia models in terms of our present discussion. The global nature of many of the processes involved in descriptions of Gaia suggest that nonadaptive selective processes may have more of a role to play than adaptive ones. We hope that distinguishing clearly the difference between adaptive and nonadaptive selection and the consequences flowing from each process will be of use to help shape and clarify future discussion of the topic.

10.14.12 The Sciences of Complexity:
Order, Complex "Adaptive" Systems, and Systems Biology

Our treatment of nonadaptive selection rightly suggests conceptual kinship with a large and growing interdisciplinary intellectual movement, the study of *complex systems* (Krakauer 2019). Likewise, the allied and partly overlapping field of *systems biology*, overwhelmingly focused on physiological, cellular, and subcellular complex systems and networks, shares a similar perspective (Boogerd et al. 2007). The sciences of complexity seek properties, mechanisms, and organizing principles common to systems with many interacting parts. Ultimately, the sciences of complexity aim to unify physical and biological systems under an overarching theory applicable to systems in general. Our aims are different: we are focused on the way that a specific kind of (nonadaptive) selective process can provide lawlike explanations for widespread biological patterns. Thus both complex systems theory and nonadaptive selection share an interest in biological systems of interacting components, stability criteria, and feedback mechanisms. But although our treatment of nonadaptive selection is consistent with many of the themes of complex systems theory, there are significant differences in goals and emphasis. Nonadaptive selection processes are not necessarily in conflict with general systems theory, but they represent a subset of factors leading to structures that have unusually strong significance for ecology and evolution, particularly at the macroscale. Comparison with two major themes of complex systems theory may illustrate the differences.

The first is the existence of *order*. Both natural and artificial systems often seem to form and ramify, developing organized structures and behaviors, without top-down controls or an overarching blueprint. Thus a primary focus of complex systems research concerns the origin and development of *order* and *complexity* that is engendered by interactions of system components, without formal direction. In such *self-organization*, elements of global order (emergent properties of the system) arise as a consequence of disordered interactions of the components (Simon 1996; Keller 2005; Boogerd et al. 2007). Examples exist from physics, chemistry, computer science, economics, linguistics, and biology. In ecology and evolution, particular attention has recently been paid to self-organization as an alternative to Darwinian natural selection (Kauffman 1993; Camazine et al. 2001; Ulanowicz 2001; Solé and Bascompte 2006). But in almost all cases, such self-ordered states are contrasted with entirely unordered states (systems exhibiting only randomness, neutrality, chaos, higher entropy states; there are many ways to describe the lack of order). From this perspective it is usually thought that a system either exhibits order or it does not. It is not always emphasized that there is a difference between the generation of nonrandom complexity in general and the kinds of organization that can give rise to meaningful mechanism-based explanations for particular biological phenomena (Keller 2005).

In contrast, the macroscale biological systems in which we are interested are already ordered and complex, and their complexity is no mystery (cf. McShea and Brandon 2010; Brandon and McShea 2020). In particular, ecological populations and communities are complex because many individuals or species come together and live in the same place. Competition for resources guarantees a degree of interaction among them. Thus, we are not particularly concerned with the origins of life, ordered complexity, and interaction per se. Rather, we are interested in the actual *selective* mechanisms—based on properties of existing biological systems—that generate the differences in stability of the target entities. In particular, we are interested in those patterns produced by nonadaptive selective processes in biology that give rise to general, lawlike explanations for biological patterns at the macroscale. For us, the alternative to nonadaptive selection is either adaptive selection or no selection at all, but not necessarily randomness.

Second, among ordered systems, complexity theory is concerned with how they work and how (as systems) they respond to changing circumstances. Complexity theory tends to regard systems as being similar to machines, with network architectures, cybernetic (feedback) mechanisms, and some way to respond to external stimuli. Frequently, the phrase "complex *adaptive* systems" is used to refer to these machinelike entities; *adaptation* is often used ambiguously but generally refers to transformational change (see section 3.1.4), analogous to growth, development, or reaction to stimuli. Frequently there is a blurring of transformational and variational processes. Further, the properties of systems that result in historical building of features that solve "challenges" or "problems" at the system level in particular environments are often vague or unspecified. However, see John Holland (1992, 1995), Steven Frank (1996), and Axel Rossberg (2013) for attempts to apply the concept of adaptation as we understand the term here to complex systems of various kinds.

Clearly, nonadaptive selection—as a process of selection on stability states of system components or system properties as a whole—is one way that systems can self-organize and can be one source of potential feedback mechanisms (Levins 1970; Levin 1998; Keller 2007). One might imagine that the field of complex systems would be a rich source of nonadaptive selection mechanisms and scenarios. However, we do not find this to be the case, at least at present. The reasons primarily reflect the different foci of the fields. For complex systems, the processes are regarded as properties of complex systems in general and are abstracted accordingly. In nonadaptive selection, the processes of interest are specifically *selective* ones that have an underlying biological mechanism, discernible at the macroscale. Thus, nonadaptive selection studies are likely to uncover causal explanations whose domains may be more restricted than those of most interest for general complex systems.

Awareness of this dynamic between the sciences of complexity and nonadaptive selection can potentially enrich both. But from our perspective in evolutionary biology and ecology, the tendency to reinterpret every successful mechanistic explanation in terms of the most general formulation possible can obscure the degree to which it is particularly useful in a specific domain.

Nevertheless, we believe that nonadaptive selection, as a type of process, may well have considerable explanatory power in the areas of complex systems theory, as well as approaches in systems biology; nonadaptive selection may appear in these places already but may not be identified as such. If, as we conjecture, nonadaptive selection is discovered to be a useful conceptual framework not just at the macroscale but at lower scales, this recognition will help to unify the biological sciences, as well as the study of nonbiological systems. Further consideration of these questions is beyond the scope of this book, but we hope that the present essay encourages others to pursue such investigations.

10.14.13 Evolutionary Constraints

There are two conceptually distinct concepts that are called "constraints" in evolution. One kind is supposed to determine the possible states of living systems; the other, to "constrain" and "guide" history down certain paths in a contingent sequence of historical states, subject to local adaptive selection. Gould (2002, 49) referred to the former as *structural constraints* and identified them as "consequences of physical principles." He added that they are "nonselective." The latter class of constraints are *historical constraints*, and Gould identified them as "channels from particular pasts." That is, they are constraints arising from the particular, contingent circumstances that a species encounters in the present, as a result of arriving at the end of a particular historical sequence of contingent states.[13]

Historical constraints have received considerable discussion in the literature. If one does not deal in some way with the historical nature of adaptations one may, in effect, over- or underestimate the contribution of current circumstances to the origin, function, and specific form of a feature hypothesized to be an adaptation (Gould and Lewontin 1978). Evolutionary biologists have long been aware of this issue and have developed many practices to ameliorate it (Mayr 1983; Orzack and Forber 2017).

In fact, local adaptive selection as a causal process is in no way constrained by history. Selection evaluates phenotypes in terms of their *current* relationship to fitness only, not in terms of their history. As has been said, there may be hierarchy in the world of natural selection, but there is no aristocracy (Damuth, 1988, quoted in Gould 2002).[14] It is, instead, the *historical sequence* that history constrains. As in all historical sequences, it is the previous results of natural selection and other historical events that provide a historical context for everything that happens later, including limits on adaptive evolution.

No matter how one tries to distinguish them from a contemporaneous adaptive selection process, there is no avoiding the fact that historical "constraints" are (mostly) products of the history of adaptive evolution within which they have arisen and thus are historically contingent in the same way that the adaptations themselves are. Thus, it may be possible to talk as if evolutionary history were a balance between adaptive selection and *historical* constraint, but it is hard to see how evolutionary history can be said to be *ultimately* constrained if both the phenotypes and the constraints are subject to evolutionary modification. Thus, the realm of adaptive evolution encompasses, creates, and "overcomes" its own constraints throughout an indefinite historical sequence of contingent circumstances. Yesterday's adaptations, or their correlates, are today's historical constraints.

Moreover, historical constraints do not constrain the future end products of long historical sequences to be either dissimilar or similar to each other. Adaptive evolution at any point in time and space is like starting out on a long trip. There is no *historical* constraint about how far one can ultimately go, or the location of one's final destination, only a constraint on where and how far one can get in a given period of time from where one is starting. Historical constraint is thus a necessary component of an adaptive theory and of any attempt to explain fully the existence of particular adaptations.

However, in this book we are not concerned with adaptive processes or constraints, but rather with the effect of nonadaptive selective processes in causing long-term, lawlike regularities. Nonadaptive selection processes give rise to states that are *possible*, but among which are differing levels of stability, based only on their intrinsic properties. So now let us revisit *structural* constraints, considered to be the ultimate boundaries of what is possible in organic form. Such constraints are largely independent of space and time and are universal within a broad realm. In this they resemble nonadaptive selection processes and the lawlike regularities that nonadaptive selection generates.

Nevertheless, traditional structural constraints and nonadaptive selection processes are obviously not the same. For example, the selective process that underlies the energy-equivalence relationship described in section 2.3 is neither based on physics nor is it nonselective. Rather, it is the *biological* properties of the assemblages that reach and maintain an equilibrium through a biologically based process of nonadaptive selection. Yet this process broadly constrains the variation we see in energy-use relationships, independently of time and space.

So nonadaptive selection processes are neither restricted to physical properties or laws, nor are they nonselective. We could consider them a kind of constraint on the possible results of evolution, akin to structural constraints as here defined. It would probably not be wrong to think of lawlike processes stemming from nonadaptive selection as some kind of evolutionary constraint. However, given that biological constraints are widely thought of as *opposing* selection, and that nonadaptive selection is in our usage primarily a biological process, it seems to be less confusing to discuss it by itself, regardless of its potential relationship to concepts of constraint.

Acknowledgments

In a long-developing project such as this, many people contribute support in various ways. We would like to thank the attendees of the workshop at the Centre Interfacultaire Bernoulli, EFFL, Lausanne, Switzerland, where we discussed "stability" issues in biology: Jon J. Borrelli, Stephano Allesina, Priyanga Amarasekare, Roger Arditi, Ivan Chase, Robert D. Holt, Dimitri O. Logofet, Mark Novak, Rudolf P. Rohr, Axel G. Rossberg, Mark Spencer, and J. Khai Tran. Substantial discussions with Stephan Munch clarified for us issues concerning chaos. Mark Colyvan, Rafael D'Andrea, and Blaire Van Valkenburgh read all or part of the manuscript. And finally, JD's wife, Susan J. Mazer, who read the manuscript several times, using both her magnificent editorial and scientific skills, generously contributed significant input to the project.

JD would also like to thank the many people over the years who, though not directly related to the book itself, have contributed through long conversation to the development of his perspective: as a thesis adviser, the late Leigh Van Valen, who never lost his openness to new ideas; the late Charles Goodnight, Jim Griesemer, I. Lorraine Heisler, Christine Janis, and Carl Simpson, for general patterns and mechanisms of evolution; Niles Eldredge, as postdoctoral adviser; and Jayanth Banavar and Amos Maritan, for collaboration on models of metabolic scaling.

LG would like to acknowledge a few people who were most influential in developing his ideas for this book but have not yet been mentioned or whose influence has been ongoing. First is Dick Lewontin, in whose lab LG worked part-time in 1976–77. It was most educational for LG to learn from Lewontin, and the collaboration was a fruitful one, too. LG's conceptual development owes much to the collaborations that started many years ago with Roger Arditi and Mark Colyvan on topics that, in retrospect, led in a straight line to nonadaptive selection, but seemed at the time to be pointing elsewhere. Last but not least, LG's wife, Lara Borrell, deserves a special mention. Lara's degree is in civil engineering, but she knows enough biology to understand the main ideas of our book. We are grateful for her inspiring suggestions for the use of abstract art for the book cover, which guided us in its final design.

Notes

Chapter 1

1. We use *adaptation* throughout this book exclusively to mean evolutionary change in a population that results from a process of selection, where the fitness values result from an interaction of the phenotype with its environment. In other words, this is ordinary Darwinian adaptive evolution by natural selection. This is the usual usage in evolutionary biology and much of ecology. In some other biological fields, as well as in a wide range of applications outside of biology, adaptation can refer to short-term, direct responses of a system to environmental stimuli, unconnected to any explicit selective process (West-Eberhard 1992). These are transformational, not selective changes (see chapter 3). In many fields, especially outside of biology, the two usages of adaptation are not always clearly distinguished, leading to potential confusion.

2. Causality is an enormous subject, as is the concept of explanation in science (Woodward 2019), and we cannot possibly address all of the unresolved issues that have long claimed the attention of both philosophers and scientists. However, we think that readers will recognize the distinction we make here between a causal mechanism and other modes of causal explanation. In a mechanistic causal explanation, under specified circumstances (and with respect to a given timescale), a particular result is inevitable. That is, once the initial conditions are set, the outcome is "algorithmic" as the mechanism "operates"; the result can be viewed as another set of initial conditions, and the mechanism can then repeat. This describes causal mechanisms such as exponential population growth or selection among organisms, from which it is possible to predict with some precision the results for the next time period (year, generation, etc.), based on current conditions. As described in the text, such extended predictions cannot necessarily be extrapolated *far* into the future, as not all of the relevant initial conditions for each iteration are likely to be predictable or constrained by the mechanism itself. In contrast, for nonadaptive selection, the relevant initial conditions are *part* of the mechanism, being the inherent

properties of the entities undergoing selection. The effect of variation in those properties on fitness does not rely on information external to the mechanism. Thus, the action of nonadaptive selection *can* be indefinitely extrapolated.

The other familiar mode of causal explanation found in many sciences—historical explanation—is very different. Purely historical "explanation" is presented in terms of a chain of antecedent states. As such, it is able to offer an account of the genesis of a unique event but is unable to serve as a basis for generalization to explain repeated patterns or to produce lawlike mechanisms. This is because it says, in effect, that in each case a unique combination of causally relevant conditions happened to coincide, and this combination caused the event (the bolide impact, the mass extinction, the origin of a new body plan, the loss of some ancestral trait value, the cladogenetic event, the specific change in gene frequency, etc.). All of those preceding conditions are part of the explanation and thus simultaneously refer to both the general nature and the uniqueness of the event. Scientific reasoning and knowledge of causal relationships among observable phenomena may play a role in explaining unique historical events, but historical explanation does not make use of a mechanism involving the event itself. There seems to be no agreed-upon way to establish the causal connections between past and present events on which historical explanation is based; the variation in scale and nature of the events to be explained is too vast. Much of the focus on explanation in sciences such as biology and geology is oriented toward historical inference (Sober 1988; Cleland 2011). That is to say, the focus is on inferring the occurrence of previous historical events in the chain leading up to the target event, which events may have influenced the target event. In this, historical inference is conceptually closely related to the growing development of "causal inference" from complex observational data (e.g., Pearl 2018). Such studies rely on statistical characterizations or correlations from which some causal influences are inferred but which nevertheless are not as yet supported by explicit causal mechanisms.

3. As used here, the *domain* is the conceptual space within which conditions exist such that a given process may and does operate. Nonadaptive selection processes always have some domain where they apply, and outside that domain conditions are such that the selection process is not possible. In other words, a domain consists of all those circumstances that are necessary for the process to be in force. For example, Darwinian adaptive natural selection is "universal," but it does not take place where there are no organisms or in the middle of the sun. Such places lack critical features for natural selection to operate and are outside its domain.

Chapter 2

1. In practice there are several different definitions of the interaction co-

efficients in use, with somewhat different implications for the measurement of their values (Novak et al. 2016). However, we are concerned here primarily with the signs of the matrix elements, which remain the same across various definitions of the interaction terms.

2. Though as an incidental effect the community as a whole may, on average, be more "stable," in the sense of not changing as frequently. As discussed in more detail later, this does not imply anything about an adaptive fit of the community to its "environment."

3. Note that this does *not* assume that species have the same extinction probabilities in the community across body sizes (although the data described later suggest that this is approximately true). It is sufficient that at any given size the rarer species at that size are more prone to extinction. Many people confuse population *density* with population *size*. For example, in conservation biology, it is often argued that species of one size or another have elevated extinction rates because of differences in population size, and this claim is backed up with density values. However, species population sizes are not measured by density (numbers per unit area), but rather by an absolute number of individuals that depends on how each species uses space. For example, elephants have very low population densities, and mice have very large densities. But mice cover much less distance during their lifetimes and regularly encounter few conspecifics relative to their density. Likewise, elephants individually cover a huge area relative to mouse individuals, and thus their population sizes can be much higher than their densities imply. So, one cannot use the measures interchangeably. What is true, however, is that among species of *similar body mass*, which likely use space very similarly, those that use less energy for their body size (low densities) *are* more likely to go locally extinct. This would be true at any body size and is the only assumption required by the nonadaptive mechanism described in the text. It is also universally agreed upon.

4. Many datasets supposedly reporting population densities have actually used proportions of different species found in a sample of a given, small, delimited area. These studies equate relative abundances in the sample to population densities, but these two metrics are generally highly distinct. Population densities may take on any positive real value. They must usually be obtained on a species-by-species basis, since it is necessary to scale the area surveyed with body size to represent fairly the density of the species where it lives. That is, the accurate estimation of the population density of a large-bodied animal requires a much larger area to be surveyed than the accurate estimation of the density of a very small one. Sample (count) data are different in that the area is standardized for all species and, significantly, no species can be observed to be present with an abundance less than "1." Such samples include insect sweep nets or fogging of tropical trees (Morse et al. 1985) and most transect samples, including Breeding Bird Survey data (Brown and Maurer 1987; Russo, Robinson and Terborgh 2003). The characteristic of such samples

is that, among the most abundant species, population abundances decline with body size as expected, but rare species all appear to have the same population abundance regardless of their body size (i.e., usually a density of "1"), as can be seen in the plots in question. The resulting plots have a "flat bottom" and have also been described as being "triangular" (Cotgreave 1993; Marquet, Navarrete, and Castilla 1995).

There was a flurry of interest in this triangular pattern in the late 1980s and 1990s, with some researchers suspecting that they had discovered a previously unknown principle of size scaling that applied not only to ecology but outside of biology as well (Gaston, Blackburn, and Lawton 1993). In fact, the flat bottom is widely seen because in any sampling procedure one cannot observe fewer than one individual. Assigning a value of "1" as the "density" of all of the rarest species in the sample generates the flat bottom across body sizes. In effect, not only is the distribution of densities truncated at a value of 1, but the "1" assigned to rare species is often not in any way a representation of the true density. For example, imagine that you sample an area of a few hectares in extent. If you detect one individual of a large, rare species in your sample, you do not know whether the next individual of that species is to be found a meter outside of the sample area, in which case the population density may be fairly high, or many kilometers away, in which case the population density would be very low.

These triangular abundance distributions from samples are no longer part of the discussion in the context of energy equivalence, but reference to this obsolete older literature sometimes resurfaces (e.g., Evans 2014).

5. Even standing crop biomass, frequently used in ecology and conservation biology to compare the status or productivity of species, is usually not an equivalent currency across species of different sizes, since a given amount of biomass metabolizes at a different rate depending upon the size of the organism to which it belongs. Standing crop biomass per unit area will necessarily vary across body size as long as population density scales with body size to an exponent other than -1, as is usually the case (Damuth 1981b). Under energy equivalence, standing crop biomass will scale positively as mass raised to the $1/4$. A population's rate of production of biomass, on the other hand, scales as does population energy use among species with similar energy-to-biomass conversion ratios (see chapter 6).

Chapter 3

1. Michael Rose and George Lauder (1996, 9) presented a vivid metaphor that underscores the general frustration of coming to grips with the concept of adaptation: "Adaptation is no longer something that can be safely assumed

by evolutionary or other biologists. Indeed, the more one examines the concept, the more it comes to resemble a newly landed fish: slippery, slimy, ob-
streperous, but glittering with potential. There it is, flapping about, but the
significance of all the commotion is not clear. Perhaps the solution of some
evolutionary biologists is best—just throw the damn thing back in the water.
But of course our authors have not chosen that course of action, and we are
left with the problem of what to do with the fish."

2. Since ours is not a book about adaptation or adaptive selection processes, we have no reason to discuss the enormous literature on mechanisms
of adaptation and the historical formation of adaptations in detail. In some
cases there may arise novel adaptive innovations that confer high individual
fitness generally, and although they still originate in a single place, they eventually spread throughout a species' range (however large in extent). Such a
spread takes time and may result in slight differences throughout the range
due to locally encountered circumstances. In other cases, the novel adaptive
trait in its natal population is almost completely prevented from spreading to
(some) other locations, either because of the complexity of the genetic milieu
within which the trait arose or by the fact that environmental conditions vary
in a nonrandom way, giving rise to cline-like geographic patterns (Kawecki
and Ebert 2004). When adaptive selection is highly constrained in this way, it
is often referred to, particularly in the population genetics modeling literature,
as *local adaptation* (Charlesworth, Nordborg, and Charlesworth 1997; Chen
et al. 2012; Sanford et al. 2003) However, there must be a continuum in the
ease by which adaptive traits may spread. Since all adaptive traits on this continuum, at the extremes and all points in between, are consistent with the way
we see adaptation, for our purposes we do not need to identify populations at
only one end of the spectrum. When we say that all adaptive selection is local,
all we are doing is reminding the reader that the result of adaptive selection
is initially specific to one local population; what happens historically to the
spread of that adaptation is ordinarily of no interest to us (i.e., when discussing
nonadaptive selection). All adaptation for us is local, in the sense that adaptive evolution evaluates fitness through interaction with the environment, but
nonadaptive selection does not. Of course, many populations, especially of organisms, must experience nonadaptive and adaptive selection simultaneously.

3. A distinction between fitness and adaptation is also critical to refuting
the charge that natural selection is merely a tautology (Brandon 1978; Bernstein et al. 1983)

4. Some accounts include four, five, or more principles. This sometimes
represents simply subdivision of one or more principles for clarity. Others
add additional principles to incorporate aspects of adaptation, in order to encompass the whole process of adaptive evolution. However, Lewontin's three
conditions (see text) enjoy wide acceptance as a general description of the

logical basis for how evolution by selection—that is, the dynamics of a selective process—works.

5. Sometimes, following Darwin, this fourth principle of adaptation is split into a Malthusian component and a competition component, yielding two principles: (1) exponential population growth, which means that not all can survive, and (2) different abilities, which mean those that do survive are a subset whose members exhibit higher fitness.

6. In practice, it can be difficult to devise an empirical definition of fitness that will be equally useful for all circumstances (Van Valen 1989). For example, in some cases, fitness is well approximated by the relative number of offspring produced. In other cases, fitness may be better estimated as the relative ability to survive a particular challenge. This flexibility of the term is not a failure of selection theory; rather, it is what makes possible the widest possible application of the concept of selection and its usefulness in both adaptive and nonadaptive cases.

7. When we say that nonadaptive selection is independent of local conditions, we do not mean that what we see as a realized outcome or standing configuration of an evolving biological system is independent of local context. In a real sense, all realized outcomes analyzed in their full complexity are context dependent and historically contingent—historical events occur only once, as Heraclitus observed. All ecological entities are subject not only to selective processes but also to all events and influences, random and otherwise, that may impinge upon them. It is only when other deterministic influences, adaptive evolution, and random processes exert sufficiently small effects that the influence of nonadaptive selective processes will dominate in the final result. This makes it clear that we expect the explanatory value of nonadaptive selection will differ among levels and types of biological systems. In some types of systems nonadaptive selection can be ignored, and nothing much will be lost. In other types, major recurrent structures depend primarily on nonadaptive selection processes.

For simplicity, we refer to "the environment" as if it were an easily characterized set of external conditions. But we mean the term in a very broad sense, including physical factors, other biological systems, and species. Different species do not even experience the physical characteristics of the environment in the same way. Nevertheless, biological systems such as organisms experience some environment. Yet nonadaptive mechanisms do not see or respond to that environment. To be more precise, we would probably have to say that the environment as we mean it is the set of historically contingent circumstances that at a given moment may impinge upon some processes of evolutionary change affecting a given biological system. Nonadaptive selective processes operate the same way regardless of such contingent context and thus exhibit great generality.

8. This concept of selection has been sufficient for us to make our arguments in this book. We recognize that certain situations investigated by evolutionary biologists are not easily represented by the conventions of the Price equation, nor by an uncomplicated distinction between selection and transmission. For example, when the mechanism of transmission is itself a character under selection, clarity may demand different model structures (Michod 2005).

9. Darwin distinguished two types of selection, natural and artificial. The two did not differ in the mechanics of selection, but rather in the source of fitness differences. In artificial selection fitnesses are imposed by human agency, and in adaptive selection they arise from interaction of phenotypes with the local environment. Likewise, we distinguish a third type of selection on the same kind of basis; nonadaptive selection is certainly natural, but the source of fitness differences is internal configuration or stability criteria that operate independently of the environment. All three types of selection can be occurring simultaneously.

10. In many transformational explanations, within and outside of biology, short-term responses to environmental stimuli are often called *adaptation*. This usage is distinctly different from the strict sense used in evolutionary biology, which we use exclusively in this book. The less-specific usage of the term is familiar to biologists (e.g., in physiology and in complex systems theory) and has frequently been remarked upon, but it nevertheless can lead to ambiguity or confusion (West-Eberhard 1992). An awareness of different researchers' goals and perspectives usually allows such usages to be distinguished in context, but substantial conceptual and practical issues follow from the choice of how selection is represented in different models (Wade and Kalisz 1990).

11. Stephen J. Gould agreed with Daniel Dennett on this point about selection but not about evolution as a whole: "I am perfectly happy to allow—indeed I do not see how anyone could deny—that natural selection, operating by its bare-bones mechanics, is algorithmic: variation proposes and selection disposes. So if natural selection builds all of evolution, without the interposition of auxiliary processes or intermediary complexities, then I suppose that evolution is algorithmic too. But—and here we encounter Dennett's disabling error once again—evolution includes so much more than natural selection that it cannot be algorithmic in Dennett's simple calculational sense" (Gould 1997a).

Of course, in this case Dennett and Gould both agreed about adaptive evolution over one generation. It is the selection process alone, not the process of adaptation, that is algorithmic. Adaptation has a historical component that depends on more than the selection mechanics. Others have seen the problem of extrapolation inherent in adaptive selection, and the potential open-endedness of evolution, as a challenge for the algorithmicity of any grand evolutionary theory (de Vladar, Santos, and Szathmáry 2017). Selection can be a fully algorithmic process, but the whole of evolution (including adaptation) cannot. See also Robert Orr (1996), who argued that adaptive evolution (which is what Dennett seems to have thought is universal) optimizes only fitness, not "Design"; thus it does not do what Dennett wants it to do.

12. An algorithmic or mechanistic explanation also contrasts with a growing research program that seeks causal inference from a statistical association of model variables with quantitative or qualitative outcomes (e.g., Pearl, Glymour, and Jewell 2016; Pearl 2018). Such analytical tools are becoming common in fields where observational data predominate. In these cases the detailed mechanisms may be unknown, or unknowable, or the potential number of interactions may be too numerous to model explicitly, but it is nevertheless of interest to discover if there is evidence of some causal relationship (direct or indirect) with a particular outcome. Does taking this drug protect against disease? Does climate change affect the frequency and intensity of wildfires? Does diversity affect speciation rate? In such cases no causal mechanism is used for explanation, but rather causal influences are inferred statistically.

13. Some philosophers still suspect that there might be fundamental laws in science, and the classical meaning of laws may retain some utility (Hoefer 2008; Pigliucci 2013). Michael Ruse (1970) argued that biology does have strict laws. Many now believe that *ceteris paribus* laws, as described in the text, are the kind of laws science actually has, and that laws in biology are no different and pose no special problems (Sober 1993, 1997; Colyvan and Ginzburg 2003; Ginzburg and Colyvan 2004).

Others believe that the only things that are to be explained in biology are, like adaptation, historically contingent, and therefore there are no biological laws of any kind (Beatty 1995; Brandon 1997). We disagree with the thrust of this argument, since we think there are general regularities in biology that are not best explained by unique, historically contingent biological events, and pursuing nonadaptive selection explanations is a good alternative.

Some philosophers, such as Robert Brandon (1997), Sandra Mitchell (2002), and Jim Woodward (2002), believe that one or more expansions of the definition of laws, beyond those listed in the text, would permit laws to involve contingent content and apply to many or most biological explanatory regularities. We suspect that this may be unnecessary, if it is allowed that explanation does not require laws, and that not every explanation traces back to one.

14. *Ceteris paribus* laws suffer from one potential difficulty, in that the complexity of biology may make it difficult to specify the range of confounding variables that have to be controlled. We argue that this is not a significant problem for nonadaptive selection explanations of macroscale patterns because there, to a first approximation, all other things *are* equal—because of the absence (or ineffectiveness) of adaptive processes at the macroscale. See section 3.3.

15. What about ecosystems? Some schemes ascend from populations to ecosystems, rather than to communities. What this does is add a new layer of complexity at the community/ecosystem level by including in the hierarchical unit aspects of the physical environment along with the biological entities. Our interest in this book is in processes that operate on or involve the biological entities at various levels, and whether or not they "see" the local environment. So for consistency, and to render our project comparable at all levels, we address only biological entities. The evolution of ecosystems raises many interesting questions concerning the way that selective processes and stability issues may involve nonbiological systems, but this topic is beyond the scope of this book.

16. This doesn't mean that the construct itself cannot be said to have been in existence over a specific time span. A *taxon* is a construct defined in such a way that it must have a beginning and—if not still extant—an end, since ancestor-descendant relationships require continuity. Some researchers have argued that there is a kind of functional, mechanistic relationship between the two hierarchies, the ecological and "genealogical," at least at some level. Eldredge made this possibility concrete by calling ecological populations *avatars* (following Damuth 1985), but calling more or less equivalent genealogical constructs at the same time "demes." The "demes" are in the genealogical hierarchy because they are ledgers that contain genetic and reproductive information needed by the avatars. He wrotes, "The genealogical hierarchy sequesters genetic information. . . . When avatars are wiped out . . . it is the [deme] that supplies the raw recruits when the area is recolonized" (Eldredge 1999, 168). But this seems a superfluous complication when talking about ecological entities. Avatars (local populations) contain all of the machinery, all of the information that they need to reproduce in the economic space of the material world. They are composed of organisms, and they metabolize, grow, and reproduce in the physical, economic world. They do not need a supply of genes from a historical construct to function, reproduce, or expand, and new avatars come from other avatars (one or many) in the world, not from a ledger of genetic information. So from the point of view of actual mechanistic, causal processes, which operate exclusively in the economic, material world, there are only ecological entities. This does not mean that historically speaking there is no connection between the hierarchies—as economic events cause change among the ecological entities, the information they possess may change in the genealogical records of the corresponding historical constructs, but there is no two-way economic exchange. Similarly, George Williams (1992b) erected a "codical" hierarchy, parallel to the material interactors, to contain nonmaterial units of "information" needed by the evolutionary process. We agree with Gould (2002, 641–43), who argued that both of these schemes are unhelpful and that they risk confounding causality and bookkeeping. A movement

toward recognizing a hierarchy of "information" instead of material objects and material, spatially identifiable systems is not viable as a source of causal interaction at any level. Species and taxa, as historical constructs or records, cannot function in selection as "information systems." Likewise, attempts to cast the two hierarchies as a single complex system must fail because there is no reason to believe that members of the two hierarchies form a system with interacting parts (see Caponi 2016).

17. We do not rely on a distinction between Richard Dawkins's "replicators" (1976, 1978) and David Hull's "interactors" (1980, 1988) when discussing selection. These two concepts define a different approach to selection than ours. We base our discussion on the Lewontin conditions and the Price equation. The replicator-interactor distinction is somewhat out of fashion, at least as a general description of selection (Nanay 2011). It was motivated to address the reductionist view that genes were the only "units" of adaptive selection (Dawkins 1976). As such these concepts are intimately involved with the workings of Darwinian adaptive evolution among organisms (and their genes) and do not help to clarify the distinction we make between adaptive and nonadaptive selection. Our conception of selection is broader and does not regard selection as always being adaptive or restricted to Darwinian natural selection. See Bence Nanay (2011) for a defense of an updated version of the replicator-interactor approach.

18. We should describe the source of variation upon which nonadaptive selection works. Daniel McShea and Robert Brandon (2010) and Brandon and McShea (2020) argued forcefully for a fundamental law of biology that they referred to as the Zero-Force Evolutionary Law (ZFEL). It states that even in the absence of any other deterministic process, all biological systems that are made up of or contain parts subject to adaptive evolution will continually generate variation and tend toward greater variety, if not complexity. Thus deterministic processes generate variation at multiple levels. Moreover, any system subject to selection of any kind is subject to drift. Drift for McShea and Brandon is not just genetic drift but is a general process and may occur at levels above and below the organismic. For nonadaptive selection, this implies that once a biological system has reached a stability endpoint or equilibrium, the process of nonadaptive selection does not necessarily end. Rather, the generality of the ZFEL and sampling phenomena such as drift suggests that to some extent this end point is always deteriorating. Even in highly selected systems exhibiting lawlike patterns, unstable variants are eventually renewed through multiple processes at various levels, including adaptive evolution at the organismic level among the species or populations making up the system, reconstitution of higher-level units following major perturbations, sampling drift at all levels, and so forth. We argue that nonadaptive selection in biological systems does not run to a stable equilibrium and simply stay there (as do

most nonbiological systems such as the solar system) because of the constant erosion of stability introduced by the ZFEL and other processes. Nonadaptive selection, like all selection, never sleeps.

19. Van Valen (1991) discussed the idea of "Biotal" evolution, by which he meant the historical explanation of changes in a wide variety of biological entities, structures, and systems, including communities and larger-scale entities. His view was that selection was always adaptive (though sometimes hard to see). However, he used the term *adaptive* in the sense of any nonrandom, deterministic change that could be explained historically, at any scale, rather than a specific mechanistic process that produces local adaptation to particular habitats. He was less interested in the difficulty of defining those habitats or shared environments per se. Eclectic, historical explanations always "work," but they do not form causal, mechanistic theories (see appendix 10.16 for more examples).

20. Perhaps the most potentially powerful way that adaptations could be formed at the community level is through a hierarchically expanded version of multilevel selection explored by David Sloan Wilson (1980, 1997) and Wilson and William Swenson (2003), highlighted by Leigh Van Valen (1980b, 1991), and demonstrated experimentally by Charles Goodnight (1990a, 1990b). This is basically a community-level version of multilevel selection within species, in which the fitness of individuals depends not (only) on the groups that they live in or experience, but also on the (individuals of) other species in the community that they share. Such a selective regime could underlie the widespread evolution of interspecific cooperative and mutualistic relationships. The objection to the evolution of cooperation at this level is similar to that concerning the evolution of altruism within a species: cheaters will always win over altruists. At the interspecific level it would work the same way. If an individual of species A does something that benefits individuals of species B, but at a cost to itself, and those individuals of species B that benefited do not immediately respond with a compensatory benefit to the original individual of species A, then how will a mutualism or cooperative relationship evolve under natural selection? In effect, members of each species that do not act to benefit the other species function as "cheaters" who may gain the benefits of the other species' actions (individually or collectively) without contributing costly fitness benefits to the other species. However, similar to the case of individual altruism, population structure (in this case consisting of community patches or subcommunities rather than infraspecific groups) would allow the occurrence of nonrandom interactions among individuals of different species, which in turn would permit the evolution of interspecific cooperation other than simple mutualisms. However, such a multilevel selection analysis would still be an analysis only of selection dynamics and not of the resulting locus of adaptation. Moreover, despite its intuitive appeal, the empirical work required

for fuller study is extremely arduous, and as yet this kind of selection dynamics has not seen much development. Furthermore, even if adaptive evolution among species populations in particular local communities commonly results in a level of general interspecific cooperation, there is little reason to expect that such evolution would result in or override the nonadaptive selection described here and the lawlike regularities that result.

Chapter 4

1. Fisher was unaware that Darwin had previously made the same argument in the first edition of *The Descent of Man*, but withdrew it from later editions (Edwards 1998).

Chapter 5

1. Of course, any compound variable consisting of the product of a physiological rate or cycle frequency (tending to scale as $-1/4$) and a longevity ("lifetime") variable (scaling as $+1/4$) will generate such a scaling invariance (Lindstedt and Calder 1981; Calder 1984; Fowler 1988; Charnov 1993). Thus, for example, in a lifetime the numbers of heartbeats, breaths, cell divisions, and potential births are all invariant with body size. Thus, the existence of an invariance does not, alone, allow one to infer the causal process determining the whole network of algebraically interlocked variables. See chapter 6 for further discussion of such networks.

2. By "fitness" the model of these researchers usually seems to mean "energetic fitness" or "adaptedness," represented by control or use of energy resources by individuals (which at equilibrium, or carrying capacity K, represents the energy used to produce each individual that makes it to the next generation). This adaptedness level seems able to evolve only when K changes (more or less by definition).

3. Different models of density-dependent population growth generally have differently shaped curves and reach the May threshold at different values of R_{max}. The actual shape of density dependence (and hence the most appropriate model) is usually not known for most natural populations, and it is likely that it varies among higher taxa and, to a lesser extent, within taxa. Our assumption is only that related species are likely to have generally similar curves, or at least that the curves' shapes do not vary consistently with body size (Fowler 1981; Sibly et al. 2005). Likewise, we take no position on how severe or unpredictable the instability must be before extinction probability is substantially elevated. All we assume is that there is some level of instability—perhaps still short of full chaos—that leads to extinction. Most or all ecological

systems likely contain the "seeds of chaos" (Berryman and Millstein 1989). Individual selection that increases R_{max} coupled with nonadaptive selection removing populations with R_{max} values that exceed the May threshold leads to an apparently "suicidal" trajectory, producing a set of species living "on the edge of chaos" (Kauffman 1993, 232, although that is not exactly what Kauffman meant by that felicitous phrase).

4. This nonadaptive selection mechanism may bring to mind Vero C. Wynne-Edwards's claims about group selection, but there are critical differences (see also 216n11). In both cases, selection among populations, for population-level properties, is seen to be significant. However, for Wynne-Edwards the selective mechanism at the higher level was the universality of an implausible mechanism of population regulation. Even more contentious was his implication that specific behaviors or traits that individuals possess are not adaptations deriving from selection on their own relative fitness, but rather must be seen as adaptations for the maintenance of a particular kind of population regulation directly selected at the higher level. We, on the other hand, think that our nonadaptive selection process involving population growth rates (not regulation) is a plausible, dynamic mechanism because of the effect of R_{max} on population-level stability, but its action is not visible at the organismic level. For us, organismic adaptations and life history strategies do not result from nonadaptive selection at the population level, nor do we regard our nonadaptive selection mechanism as constraining in any way the traits that might evolve in populations by organismic selection. There are nevertheless negative consequences for the stability of populations with high growth rates, implying that some combinations of organismic life history traits may be less often observed than others across an ensemble of species.

Chapter 6

1. The fact that we do not take a position on which of the two models of metabolic scaling is more effective does not mean that we are personally neutral. JD has been a collaborator with Jayanth Banavar and colleagues on the development of the transport network models. See Banavar et al. (2014) for a resolution of the differences between the Rubner and transportation models. LG, on the other hand, maintains a respectful skepticism.

Chapter 7

1. *Interference* refers to all of those specific biological processes that lead to a decrease in per capita predator consumption with an increase in predator relative abundance. This includes increases in frequency or severity of

direct agonistic interactions among predators. The term also includes indirect effects, such as increased difficulty of locating and capturing prey because of the consumption activities of conspecifics. Moreover, it can include increased avoidance or defensive behavior on the part of the prey in response to greater perceived predator abundance. The parameter m, described a little later in the text, represents "interference" in this general sense, making no assumptions about the detailed biological mechanisms causing it.

2. This assumption is equivalent to saying that predator populations do not experience density-dependent growth and regulation. Recall that in the previous section we argued that for single-species populations, density dependence is the norm, rejecting alternatives (where population members do not interfere with each other) as structurally unstable. Thus, prey dependence implies that when given a constant resource, predator populations will be density independent and either grow to infinity or decline to zero, but when paired with their prey, predators will be transformed into populations that will have a joint equilibrium with their prey. By contrast, under ratio dependence both predator and prey exhibit density-dependent growth.

3. Although Michael Rosenzweig (1971) saw no easy way that predator-prey pairs could avoid the paradox of enrichment, he did think that there might be a counteracting factor to the instability caused by the evolution of greater predator effectiveness in systems with m near 0. Rosenzweig (1973) postulated that the destabilizing effect caused by predator evolution was blunted to some degree by an evolutionary arms race between predators and prey: predator evolutionary advances are followed by prey evolution that tends to bring the system back into stability. Such finely balanced coevolution is far from inevitable, however (Abrams 1986). Rosenzweig (1973) recognized the tenuousness of his coevolutionary postulate by arguing—in good nonadaptive selection fashion—that systems in which coevolution had failed to counteract predator evolution for greater efficiency were unstable and thus had been removed from our view. Since this destabilizing effect of predator improvement is found only in models in which there is no mutual interference among predators (such as Lotka-Volterra or Rosenzweig-MacArthur; see Rosenzweig and MacArthur 1968), this amounts to nonadaptive selection against systems where m is close to 0. We agree completely with Rosenzweig and point out that in addition to his nonadaptive selective disadvantage, the other factors that also lead to instability at the prey-dependent extreme (such as the paradox of enrichment) make things even worse for the stability of predator-prey pairs where m is very close to 0. The result of all of these factors together is strong nonadaptive selection against such systems, which is why we less often see them in nature (see figure 7.2).

4. It is possible that the pattern of successive rarity of populations exhibiting $m > 1$ could result from adaptive evolution; consumer populations at high m *might* sometimes evolve toward larger and more stable populations by de-

creasing *m*. But this scenario depends on everything being "just right." Simple Darwinian selection among individuals in such a case would require that consumer individuals that interacted with one another less strongly would have higher relative fitness, which cannot be a general case. Probably a more effective process for causing directional selection on consumers to decrease high *m* would be some form of multilevel selection. However, we emphasize that our nonadaptive selection mechanism requires no evolution of individual populations at all and is quite general.

5. May (1973a, 174) remarked on the apparent mismatch between theory and observation. He suggested that the focus of future research should be "to elucidate the devious strategies making for stability in enduring natural systems." This expression suggests that he thought that the answer to why actual diverse communities are stable lies with the adaptations of individual species. It is up to them to somehow contrive (through adaptive coevolution?) the possession of specific properties and strategies that make communities stable. This perspective has guided much research on the (joint) species properties that would confer stability under various circumstances, and the circumstances that might cause interacting species to evolve particular relationships. We describe this approach as "quantitative stability" in section 7.4.1. However, this outlook discounts the possible influence of general, nonadaptive selection processes in influencing what we ordinarily see. May's result is based on his assumption of interaction symmetry. It is now increasingly clear that relaxing that assumption allows for strong one-sided asymmetric interaction matrices that match widespread empirical patterns.

6. The effects of predator and prey numbers on each other's population dynamics are likely asymmetrical in a wide variety of cases, as under ratio dependence, low variation in predator numbers has little effect on prey population growth. This asymmetry was expressed by Rolf Peterson (2013) as, "prey matter to predator populations far more than vice versa." Arditi and Ginzburg (2012) called this asymmetric situation "donor control," a term coined by Stuart Pimm (1982) for a more restricted special case. Pimm applied the term to situations in which the predators were consuming only dead or dying prey, ones that were about to die "anyway." Clearly, consumption of these resources could not affect the dynamics of prey population growth. Arditi and Ginzburg (2012) showed that this version of donor control is a special case of the expanded, ratio-dependent version described in their book. Pimm's original version generates asymmetry regardless of population model.

7. See section 2.2 and appendix 10.12 for the representation of a web or network as a matrix of interaction coefficients. Ecologists have used a number of slightly dissimilar types of community matrices (Novak et al. 2016). They differ in the definition of the interaction term; the response variable can be in terms of absolute growth rate, per capita growth rate, or more complex

constructions. These alternative choices will not concern us. Here, as before, we are mostly talking about the signs of the matrix elements and the asymmetry of their magnitudes, whose effects remain the same across various definitions of the interaction terms.

8. This felicitous term introduced by Ron Milo et al. (2002) is borrowed from music and describes sets of commonly recurring network structures (such as community modules) as *motifs*. Motifs form classes of networks that share sets of basic structural forms, and different motifs may characterize networks in different fields (including those outside of biology, such as studies of the internet), or the same motifs may span a variety of fields. Continuing the musical analogy, some motifs are repeated more frequently or are heard more clearly than others. Some network motifs may be so common that they are regarded as fundamental observations or concepts in their field.

9. If symmetry is assumed to be the general case, there would be no reason to investigate the triangular matrix, in which interactions are strongly asymmetric. Interestingly, May (1973b), when introducing qualitative stability to ecology, did not interpret the triangular case biologically, though he was certainly aware of it.

10. "Because a frequent outcome of coevolution is the loss of species[,] . . . [c]oevolution may, so to speak, prune the strongly interacting species from the original community, leaving a community that is less diverse and as weak interactions among most remaining members. . . . Thus, it is not by 'molding' the interactions, but by causing extinctions, that coevolution may ultimately generate stable communities" (Roughgarden 1983, 62). Of course, we would say that it is the nonadaptive selection against unstably interacting species, whether those extinctions are caused by coevolution or not, that generates more stable communities over time.

Chapter 8

1. *Macroevolution* has no generally accepted usage (Eldredge 1989). Definitions vary, from essentially forming a subset of microevolutionary processes that underlie evolution of major differences among taxa (Dobzhansky 1970; Levinton 2001), to any evolutionary processes that act on geologic timescales (Gingerich 1987), to evolutionary phenomena occurring primarily "above the species level," which may be the most common current usage, although not without its own ambiguities (Simpson 1944; Stanley 1975, 1979).

Since the neo-Darwinian synthesis in the mid-twentieth century, macroevolutionary studies have overwhelmingly focused on the description of historical patterns, including long-term rates of change, diversification of higher taxa, major events in life's history, and the origin and fate of evolutionary novelties (Simpson 1944, 1953). Such research explains historical patterns primarily by delineating sequences of events, rather than by using mechanisms (Currie 2019). History can be a satisfying and effective explanatory mode, but ultimately, each outcome appears as the result of a unique sequence of events. Mechanism does not often enter the picture; historical explanation has no need for it. And it doesn't matter that most of the time in macroevolution we are talking about taxa (historical constructs) as opposed to typical ecological-based entities, since history can be written using any entities that exist in time.

The generalizations that have been advanced about macroevolutionary history are usually statements about classes of unique historical events with some aspects that appear to recur (e.g., Raup and Sepkoski 1984; Gould 1988, 1997b; Jablonski 1997; Marshall 2017).

Chapter 9

1. Dawkins (1976, 13–15) expressed this idea as follows:

> Darwin's "survival of the fittest" is really a special case of a more general law of survival of the stable. The universe is populated by stable things. A stable thing is a collection of atoms which is permanent enough or common enough to deserve a name. It may be a unique collection of atoms, such as the Matterhorn, which lasts long enough to be worth naming. Or it may be a class of entities, such as rain drops, which come into existence at a sufficiently high rate to deserve a collective name, even if any one of them is short-lived. The things which we see around us, and which we think of as needing explanation—rocks, galaxies, ocean waves—are all, to a greater or lesser extent, stable patterns of atoms. Soap bubbles tend to be spherical because this is a stable configuration for thin films filled with gas. In a spacecraft, water is also stable in spherical globules, but on earth, where there is gravity, the stable surface for standing water is flat and horizontal. Salt crystals tend to be cubes because this is a stable way of packing sodium and chloride ions together. In the sun the simplest atoms of all, hydrogen atoms, are fusing to form helium atoms, because in the conditions which prevail there the helium configuration is more stable. Other even more complex atoms are being formed in stars all over the universe, and were formed in the "big bang" which, according to the prevailing theory, initiated the universe. This is originally where the elements on our world came from.

2. Memorably quoted by Fisher (1930).

Chapter 10

1. Although knowledge of them was valuable in assessing the status of putative adaptations, disproportionate change in a character might just reflect a correlated response to less obvious selection on another linked character.

2. Sometimes called the Price theorem.

3. The equation does not generate its before-and-after comparison by simply taking the means of z in the two populations, as much as it "generates" the second population from information about the individuals in the first population. In such a way it keeps changes in the frequency of individuals with different z values separate from transformational changes of z in those individuals (or their offspring).

Note also that for evolution to occur as a result of selection, there is no requirement for specific mechanisms of reproduction. All that the Price equation requires is that there be some resemblance between parent and offspring. So a statistically consistent like begets like of any kind will do.

4. The covariance relationship was independently discovered by Alan Robertson (1966), by Ching Chun Li (1967), and finally by George R. Price (1970).

5. Strictly speaking, the effects of random drift also would appear in this term, in the form of random variation in the fitnesses assigned to a given value of z. In theoretical discussions drift is frequently ignored (see Rice 2004).

6. Derivations from the Price equation can become quite involved. Not all are equally valuable in representing particular processes, though they are all mathematically equivalent descriptions of the overall effect of selection, as described by the original equation. For example, *multilevel selection* refers to an approach to analyzing selection in hierarchically structured populations in which individuals interact nonrandomly with neighbors, relatives, social group members, or subpopulations, and these interactions generate effects upon individual fitnesses (Wade 2016). Price's (1972) own representation of multilevel selection separated the effect of selection on individuals into within- and among-group components and has been influential because of its relative simplicity. Because it is a mathematically correct partition, it describes accurately the change of the mean value of an individual character, across the whole population. As such it is directly useful for certain general questions and, particularly, for analysis of theoretical models in which multilevel selection processes have been intentionally included as a component (Wade 1985; Sober and Wilson 1998). However, Price's partition does a poor job of isolating the group-level effects on fitness from other contributions to fitness differences among groups that may occur in nature (Heisler and Damuth 1987; Goodnight, Schwartz, and Stevens 1992; Okasha 2006). To answer the question, "Are there group effects on fitness over and above variation in fitness of individuals?," one requires a contextual analysis, a regression approach also

derivable from the Price equation. In fact, contextual analysis is a form of multilevel selection that can most easily be seen to be mathematically equivalent to inclusive fitness approaches, which are also derivable from the Price equation (Goodnight 2013; see also Okasha 2016). NB: In the multilevel selection literature, Price's multilevel partition is often referred to simply as "the Price equation," though it is only one possible partition of the Price equation.

7. This does not mean that the two approaches (and others such as evolutionary game theory or identifying the evolutionarily stable strategy) are equally useful for addressing specific questions, nor do they necessarily provide equally natural ways to extend models to more complex genetical models. The approaches can be considered to be complementary.

8. Elliot Sober and David S. Wilson (2011, 465), in a critique of Andy Gardner and Alan Grafen (2009), argued for a simple rule of thumb:

> What is good for the individual can conflict with what is good for the group. The concept of adaptation should reflect this fact. Rather than use "individual adaptation" as an all-encompassing label that is defined so that it applies to all adaptations regardless of whether they evolve by group or individual selection (or any mixture thereof), we think it more useful to use "group adaptation" to label traits that evolved when group selection dominated the selection process and "individual adaptation" to label traits that evolved when individual selection was in the driver's seat. Why have two labels if one of them applies no matter what?

Sober's solution seems to us a convention for convention's sake, not a fully formed attempt to establish criteria for detecting adaptation at a level. It would seem to work only when selection at the two levels was opposing and would not illuminate the case of our hypothetical example. Most importantly, such a resolution still misses the point that the selection dynamics alone cannot meaningfully define what an adaptation is or whom it may benefit.

9. In some cases we can discern nonadaptive selection at work. For example, in a subfield of language evolution, evolutionary phonology (Blevins 2006), certain combinations of phonetic elements are more likely, and specific changes from one to another are more likely to be observed. This is because of physical characteristics of the human speech organs, innate information-processing abilities of the human senses and brain, and a variety of other possible influences (including some that are apparently transformational). That is to say, such changes result largely from selection based on quasi-universal intrinsic human properties. The kinds of phonetic changes observed neither make the language a better one (i.e., increase the effectiveness of a language

as a communication medium) nor increase the effectiveness of a given speaker. What Blevins attributes to either evolutionary "adaptation" or "self-organization" seems more simply described as nonadaptive selection.

10. In recent years it has become common, particularly when discussing evolutionary biology for the public, to define scientific theories as grand, comprehensive conceptual structures supported by enormous amounts of evidence (e.g., National Academy of Sciences 2008, 11). This is certainly true in the context of the theoretical underpinnings of evolution by natural selection. Such a concept of theories serves a pedagogic purpose in decisively counteracting the vernacular sense of "theory" as "an unfounded speculation." However, it would be a distortion to conclude from this stereotype that all theories are well-established, fully articulated, formalized, and of broad scope. In particular, in daily practice biologists seldom make use of or make reference to, for example, "the" theory of evolution, in its entirety, but rather concern themselves with investigating topics of much more limited scope, related to subtheories within the overarching framework of evolutionary theory (see Getz 2006). Nowhere is there a list of all the theories, laws, and models identified or used in a scientific community, nor is it likely there can be. Such a list would contain almost all of the conceptual, general causal explanations advanced by scientists, some well-established, others as yet incompletely specified and with only meager (but perhaps tantalizing) empirical support (Scheiner 2010). The most well-established theories of the broadest scope are unlikely to be overturned on new evidence, but at the other end of the scale, where scientific work is focused and theory construction is occurring, the situation is more fluid. There appears to be no reason that theories or other conceptual tools should be limited in number or that they should require a particular breadth of scope.

11. Wynne-Edwards's theory needed a somewhat vaguely specified group selection process to produce individual behaviors which, in turn, would guarantee an equilibrium population size *below* what we now define as carrying capacity (K). Wynne-Edwards thought such populations would be "better" populations because they did not exhibit high rates of juvenile mortality or individuals that were sick or starving. In his view, nature inevitably led to the minimization of suffering among organisms. So fundamentally it was not so much an evident adaptation of groups that Wynne-Edwards thought he detected, but a human social value that he imposed on the workings of nature, which had supposedly selected for certain adaptations of individuals to benefit all the members of a group. This is particularly clear in his description of the implications of his theory for human society. Wynne-Edwards believed that like all other species, humans once had had the kind of behavioral mechanisms in place to keep their populations below K. But in modern times these evolved behaviors were now ineffective, as a result of rapid technological progress and cultural change (Wynne-Edwards 1965). The result was poverty,

disease, hunger, and runaway population growth, which would be disastrous unless society took the step of instituting new "conventions" that would revoke the right to reproduce for some individuals.

By 1986 Wynne-Edwards (1986, 357–58) had convinced himself that group selection always trumps individual selection, and that individual selection is of little consequence except to identify individuals of high "worth" to the group:

> Individual and group selection can be successfully combined because individual selection is only permitted as long as it benefits the group. (If it were to persist, notwithstanding, it would soon reduce the existing fitness levels and become self-defeating.) It is used solely for the purpose of showing which individual animals are in the most viable and socially dominant class, and then according them the rights to feed and breed. The actual breeding density, and the due quota of progeny, are regulated by group-selected programs; and individual selection itself is largely artificial, in the sense that infraspecific competition is ruled by group-selected conventions. Only the extrinsic agents that deal injury and death impinge in the raw, and even these may fall heaviest on social outcasts.

The repeated references to "most viable and socially dominant class" versus the "social outcasts" who should be prevented from breeding freely are (and should be) disquieting to the modern ear.

In Wynne-Edwards's day human overpopulation was a widespread concern, and our understanding of the mechanisms of natural population regulation was not yet resolved (McLaren 1971). Today density-dependent regulation is not thought to depend on anything but the eventual scarcity of resources, as Malthus, Darwin, David Lack (1954), John Maynard Smith (1964), and others claimed, and populations do not have to be regulated below K in order to be viable. The idea that nature avoids suffering seems unsupportable to us; it doesn't matter to nature how many baby turtles die on the beach as long as enough of them make it to the sea. Wynne-Edwards was a great naturalist, but his evident love of nature seems to have led him to endow it with a degree of benevolence that it does not possess.

12. Lovelock was primarily a chemist and engineer by background. It is clear that in likening Gaia to an organism, he was focusing on the cybernetic (feedback) mechanisms that characterize living systems and maintain homeostasis. In this sense he saw both Gaia and organisms as akin to machines, and it is this similarity that drove the metaphor. The evolutionary processes underlying the origin and elaboration of the biological components were of secondary interest to their current functioning (see Ruse 2013).

13. Evolutionary constraints are often defined as any of a wide variety of factors that oppose the effectiveness of adaptive selection across one generation (Antonovics and van Tinderen 1991; Björklund 1996). As such, they are often seen as constraints on the adaptational process alone and not upon long-term history. Nevertheless, these constraints belong to the historical constraint category, since they themselves are contingent circumstances produced by history. The short timescale of such constraints just means that they are being recognized only at the end point of the historical sequence to which they belong.

14. In his magnum opus, Gould (2002, 672) quoted with approval a passage including this remark about aristocracy, which he attributed to John Damuth and I. Lorraine Heisler (1988). However, this text does not appear in Damuth and Heisler (1988), though it is consistent with the argument made in that article. Instead, Gould was quoting an unpublished manuscript by Damuth, a position paper entitled "Higher-Level and Multilevel Selection," circulated privately for an NSF-supported conference, Foundations of Evolutionary Biology, held at Ohio State University in July 1988 (Award 8720624, principal investigators David Hull and Michael J. Wade, and hosted by Sandra Mitchell).

References

Aarssen, L. 1988. "'Pecking Order' of Four Plant Species from Pastures of Different Ages." *Oikos* 51:3–12.

Abrams, P. A. 1986. "Is Predator-Prey Coevolution an Arms Race?" *Trends in Ecology and Evolution* 1:108–10.

Abrams, P. A., and L. R. Ginzburg. 2000. "The Nature of Predation: Prey Dependent, Ratio Dependent or Neither?" *Trends in Ecology and Evolution* 15:337–41.

Agrawal, A. A. 2017. "Toward a Predictive Framework for Convergent Evolution: Integrating Natural History, Genetic Mechanisms, and Consequences for the Diversity of Life." *American Naturalist* 190 (Supplement): S1–S12.

Allen, T. E. H., and T. B. Starr. 1982. *Hierarchy: Perspectives for Ecological Complexity*. Chicago: University of Chicago Press.

Allesina, S., and J. M. Levine. 2011. "A Competitive Network Theory of Species Diversity." *Proceedings of the National Academy of Sciences, USA* 108:5638–42.

Allesina, S., and M. Pascual. 2008. "Network Structure, Predator-Prey Modules, and Stability in Large Food Webs." *Theoretical Ecology* 1:55–64. https://doi.org/10.1007/S12080-007-0007-8.

Allmon, W. D. 1992. "A Causal Analysis of Stages in Allopatric Speciation." In *Oxford Surveys in Evolutionary Biology*, vol. 8, edited by D. Futuyma and J. Antonovics, 219–57. New York: Oxford University Press.

Amarakesare, P. 2022. "Ecological Constraints on the Evolution of Consumer Functional Responses." *Frontiers in Ecology and Evolution* 10:836644. https://doi.org/10.3389/Fevo.2022.836644.

Amundson, R. 1996. "Historical Development of the Concept of Adaptation." In *Adaptation*, edited by M. E. Rose and G. V. Lauder, 11–53. San Diego, CA: Academic Press.

Andrewartha, H. G., and L. C. Birch. 1954. *The Distribution and Abundance of Animals*. Chicago: University of Chicago Press.

Antonovics, J., and P. H. Van Tinderen. 1991. "Ontoecogenophyloconstraints? The Chaos of Constraint Terminology." *Trends in Ecology & Evolution* 6:166–68.

Arditi, R., and H. R. Akçakaya. 1990. "Underestimation of Mutual Interference of Predators." *Oecologia* 83:358–61.

Arditi, R., J.-M. Callois, Y. Tyutyunov, and C. Jost. 2004. "Does Mutual Interference Always Stabilize Predator-Prey Dynamics? A Comparison of Models." *Comptes Rendus Biologies* 327:1037–57. https://doi.org/10.1016/J.Crvi.2004.06.007.

Arditi, R., and L. R. Ginzburg. 1989. "Coupling in Predator-Prey Dynamics: Ratio-Dependence." *Journal of Theoretical Biology* 139:311–26.

———. 2012. *How Species Interact: Altering the Standard View on Trophic Ecology.* Oxford: Oxford University Press.

Aristotle. 1930. *Physics.* II.8. Translated by R. P. Hardie and R. K. Gaye. Oxford: Clarendon Press.

Arneberg, P., A. Skorping, and A. F. Read. 1998. "Parasite Abundance, Body Size, Life Histories, and the Energetic Equivalence Rule." *American Naturalist* 151: 497–513.

Arnold, A. J., and K. Fristrup. 1982. "The Theory of Evolution by Natural Selection: A Hierarchical Expansion." *Paleobiology* 8:113–29.

Arnold, S. J., and M. J. Wade. 1984. "On the Measurement of Natural and Sexual Selection: Applications." *Evolution* 38:720–34.

Arthur, W. 1997. *The Origin of Animal Body Plans: A Study in Evolutionary Developmental Biology.* Cambridge: Cambridge University Press.

Arthur, W. 2023. *Understanding Life in the Universe.* Cambridge: Cambridge University Press.

Atanasov, A. T. 2005. "The Linear Allometric Relationship between Total Metabolic Enegy per Life Span and Body Mass of Poikilothermic Animals." *Biosystems* 82:137–42.

———. 2007. "The Linear Allometric Relationship between Total Metabolic Energy per Life Span and Body Mass of Mammals." *Biosystems* 90:224–33.

———. 2012. "Allometric Scaling of Total Metabolic Energy per Lifespan in Living Organisms." *Trakia Journal of Sciences* 10:1–14.

Austad, S. N. 2010. "Animal Size, Metabolic Rate, and Survival, among and within Species." In *The Comparative Biology of Ageing,* edited by N. S. Wolf, 27–41. Dordrecht: Springer.

Austad, S. N., and K. E. Fischer. 1991. "Mammalian Ageing, Metabolism, and Ecology: Evidence from the Bats and Marsupials." *Journal of Gerontology* 46:B47–B53.

Balisi, M., and B. Van Valkenburgh. 2020. "Iterative Evolution of Large-Bodied Hypercarnivory in Canids Benefits Species but Not Clades." *Communications Biology* 3 (461): 1–9. https://doi.org/10.1038/S42003-020-01193-9.

Banavar, J. R., T. J. Cooke, A. Rinaldo, and A. Maritan. 2014. "Form, Function and Evolution of Living Organisms." *Proceedings of the National Academy of Sciences, USA* 111:3332–37.

Banavar, J. R., J. Damuth, A. Maritan, and A. Rinaldo. 2002. "Supply-Demand Balance and Metabolic Scaling." *Proceedings of the National Academy of Science, USA* 99:10506–9.

Banavar, J. R., M. E. Moses, J. H. Brown, J. Damuth, A. Rinaldo, R. M. Sibley, and A. Maritan. 2010. "A General Basis for Quarter-Power Scaling in Animals." *Proceedings of the National Academy of Science, USA* 107: 15816–20.

Banse, K., and S. Mosher. 1980. "Adult Body Mass and Annual Production/Biomass Relationships of Field Populations." *Ecological Monographs* 50:355–79.

Barabás, G., M. J. Michalska-Smith, and S. Allesina. 2017. "Self-Regulation and the Stability of Large Ecological Networks." *Nature Ecology & Evolution* 1:1870–75. https://doi.org/10.1038/S41559-017-0357-6.

Barnes, D. K. A. 2002. "Polarization of Competition Increases with Latitude." *Proceedings of the Royal Society of London B* 269:2061–69. https://doi.org/10.1098/Rspb.2002.2105.

Barraclough, T. G., A. P. Vogler, and P. H. Harvey. 1998. "Revealing the Factors That Promote Speciation." *Philosophical Transactions of the Royal Society of London, Series B, Biological Sciences* 353:241–49.

Bascompte, J. 2009. "Disentangling the Web of Life." *Science* 325:416–18. https://doi.org/10.1126/Science.1170749.

Bascompte, J., and P. Jordano. 2007. "Plant-Animal Mutualistic Networks: The Architecture of Biodiversity." *Annual Review of Ecology, Evolution and Systematics* 38:567–93.

Bascompte, J., P. Jordano, C. J. Melián, and J. M. Olesen. 2003. "The Nested Assembly of Plant-Animal Mutualistic Networks." *Proceedings of the National Academy of Sciences, USA* 100:9383–87. https://doi.org/10.1073?pnas.1633576100.

Bascompte, J., P. Jordano, and J. M. Olesen. 2006. "Asymmetric Coevolutionary Networks Facilitate Biodiversity Maintenance." *Science* 312:431–33.

Beatty, J. 1995. "The Evolutionary Contingency Hypothesis." In *Concepts, Theories and Rationality in the Biological Sciences*, edited by G. Wolters and J. G. Lennox, 45–81. Pittsburgh: University of Pittsburgh Press.

Beck, J., L. Ballesteros-Mejia, C. M. Buchmann, J. Denglar, S. A. Fritz, B. Gruber, C. Hof, F. Jansen, S. Knapp, H. Kreft, A.-K. Schneider, M. Winter, and C. F. Dormann. 2012. "What's on the Horizon for Macroecology?" *Ecography* 35:673–83. https://doi.org/10.1111/J.1600-0587.2012.07364.x.

Begon, M., and C. R. Townsend. 2021. *Ecology: From Individuals to Ecosystems.* 5th ed. Oxford: John Wiley & Sons.

Behrensmeyer, A. K., J. D. Damuth, W. A. Dimichele, R. Potts, H.-D. Sues, and S. L. Wing, eds. 1992. *Terrestrial Ecosystems through Time: Evolutionary Paleoecology of Terrestrial Plants and Animals.* Chicago: University of Chicago Press.

Béland, P., and D. A. Russell. 1980. "Dinosaur Metabolism and Predator/Prey Ratios in the Fossil Record." In *A Cold Look at the Warm-Blooded Dinosaurs*, edited by R. D. K. Thomas and E. C. Olson, 85–102. Boulder, CO: Westview Press.

Berlow, E. L., A. Neutel, J. E. Cohen, P. C. De Ruiter, B. Ebenman, M. Emmerson, J. W. Fox, V. A. A. Jansen, J. I. Jones, G. D. Kokkoris, D. O. Logofet, A. J. Mckane, J. M. Montoya, and O. Petchey. 2004. "Interaction Strengths in Food Webs: Issues and Opportunities." *Journal of Animal Ecology* 73:585–98.

Bernstein, H., H. C. Byerly, F. A. Hopf, R. A. Michod, and G. K. Vemulapalli. 1983. "The Darwinian Dynamic." *Quarterly Review of Biology* 58:185–207.

Berryman, A. A., and J. A. Millstein. 1989. "Are Ecological Systems Chaotic—and If Not, Why Not?" *Trends in Ecology and Evolution* 4:26–28.

Birch, J. 2019. "Are Kin and Group Selection Rivals or Friends?" *Current Biology* 29:R425–R473.

Björklund, M. 1996. "The Importance of Evolutionary Constraints in Ecological Time Scales." *Evolutionary Ecology* 10:423–31.

Black, A. J., P. Bourrat, and P. B. Rainey. 2020. "Ecological Scaffolding and the Evolution of Individuality." *Nature Ecology & Evolution* 4:426–36. https://doi.org/0 .1016/J.Shpsc.2014.10.006.

Blevins, J. 2006. "A Theoretical Synopsis of Evolutionary Phonology." *Theoretical Linguistics* 32:117–66.

Blueweiss, L., L. Fox, V. Kudzma, D. Nakashima, R. Peters, and S. Sams. 1978. "Relationships between Body Size and Some Life History Parameters." *Oecologia* 37:257–72.

Bohlin, T., C. Dellefors, U. Faremo, and A. Johlander. 1994. "The Energetic Equivalence Hypothesis and the Relation between Population Density and Body Size in Stream-Living Salmonids." *American Naturalist* 143:478–93.

Bolnick, D. I., P. Amarasekare, M. S. Araújo, R. Bürger, J. M. Levine, M. Novak, V. H. W. Rudolf, S. J. Schreiber, M. C. Urban, and D. A. Vasseur. 2011. "Why Intraspecific Trait Variation Matters in Community Ecology." *Trends in Ecology and Evolution* 26:183–92.

Boogerd, F. C., F. J. Bruggeman, J.-H. S. Hofmeyr, and H. V. Westerhoff, Eds. 2007. *Systems Biology: Philosophical Foundations.* Amsterdam: Elsevier.

Borrelli, J. J. 2015. "Selection against Instability: Stable Subgraphs Are Most Frequent in Empirical Food Webs." *Oikos* 124:1583–88. https://doi.org/10.1111/Oik .02176.

Borrelli, J. J., S. Allesina, P. Amarasekare, R. Arditi, I. Chase, J. Damuth, R. D. Holt, D. O. Logofet, M. Novak, R. P. Rohr, A. G. Rossberg, M. Spencer, J. K. Tran, and L. R. Ginzburg. 2015. "Selection on Stability across Ecological Scales." *Trends in Ecology and Evolution* 30:417–25.

Borrelli, J. J., and L. R. Ginzburg. 2014. "Why There Are So Few Trophic Levels: Selection against Instability Explains the Pattern." *Food Webs* 1:10–17.

Bouchard, F. 2008. "Causal Processes, Fitness, and the Differential Persistence of Lineages." *Philosophy of Science* 75:560–70.

———. 2011. "Darwinism without Populations: A More Inclusive Understanding of the 'Survival of the Fittest.'" *Studies in History and Philosophy of Biological and Biomedical Sciences* 42:106–14. https://doi.org/10.1016/J.Shpsc.2010.11.002.

Bouchard, F., and P. Huneman, eds. 2013. *From Groups to Individuals: Evolution and Emerging Individuality.* Cambridge, MA: MIT Press.

Bourrat, P. 2014. "From Survivors to Replicators: Evolution by Natural Selection Revisited." *Biology and Philosophy* 29:517–38.

Boyd, R., and P. J. Richerson. 1985. *Culture and the Evolutionary Process*. Chicago: University of Chicago Press.

Brady, S. P., D. I. Bolnick, R. D. H. Barrett, L. Chapman, E. Crispo, A. M. Derry, C. G. Eckert, D. J. Fraser, G. F. Fussmann, A. Gonzalez, F. Guichard, T. Lamy, J. Lane, A. G. Mcadam, A. E. M. Newman, A. Paccard, B. Robertson, G. Rolshausen, P. M. Schulte, A. M. Simons, M. Vellend, and A. Hendry. 2019. "Understanding Maladaptation by Uniting Ecological and Evolutionary Perspectives." *American Naturalist* 194:495–515.

Brandon, R., and D. W. McShea. 2020. *The Missing Two-Thirds of Evolutionary Theory*. Cambridge: Cambridge University Press.

Brandon, R. N. 1978. "Adaptation and Evolutionary Theory." *Studies in the History and Philosophy of Science* 9:181–206.

———. 1990. *Adaptation and Environment*. Princeton, NJ: Princeton University Press.

———. 1997. "Does Biology Have Laws? The Experimental Evidence." *Philosophy of Science* 64: S444–S457.

Brett, C. D., L. C. Ivany, and K. M. Schopf. 1996. "Coordinated Stasis: An Overview." *Palaeogeography, Palaeoclimatology, Palaeoecology* 127:1–20.

Briand, F., and J. E. Cohen. 1987. "Environmental Correlates of Food Chain Length." *Science* 238:956–60.

Brose, U., T. Jonsson, E. L. Berlow, P. Warren, C. Banasek-Richter, L.-F. Bersier, J. F. Blanchard, T. Brey, S. R. Carpenter, M.-F. C. Blandenier, L. Cushing, H. A. Dawah, T. Dell, F. Edwards, S. Harper-Smith, U. Jacob, M. E. Ledger, N. D. Martinez, J. Memmott, K. Mintenbeck, J. K. Pinnegar, B. C. Rall, T. S. Rayner, D. C. Reuman, L. Ruess, W. Ulrich, R. J. Williams, G. Woodward, and J. E. Cohen. 2006. "Consumer-Resource Body-Size Relationships in Natural Food Webs." *Ecology* 87:2411–17.

Brose, U. 2010. "Body-Mass Constraints on Foraging Behaviour Determine Population and Food-Web Dynamics." *Functional Ecology* 24:28–34. https://doi.org /10.1111/J.1365-2435.2009.01618.x.

Brose, U., P. Archambault, A. D. Barnes, L. Bersier, T. Boy, J. Canning-Clode, E. Conti, M. Dias, C. Digel, A. Dissanayake, A. A. V. Flores, K. Fussmann, B. Gauzens, C. Gray, J. Häussler, M. R. Hirt, U. Jacob, M. Jochum, S. Kéfi, O. Mclaughlin, M. M. Macpherson, E. Latz, K. Layer-Dobra, P. Legagneux, Y. LI, C. Madeira, N. D. Martinez, V. Mendonça, C. Mulder, S. A. Navarrete, E. J. O'Gorman, D. Ott, J. Paula, D. Perkins, D. Piechnik, I. Pokrovsky, D. Raffaelli, B. C. Rall, B. Rosenbaum, R. Ryser, A. Silva, E. H. Sohlström, N. Sokolova, M. S. A. Thompson, R. M. Thompson, F. Vermandele, C. Vinagre, S. Wang, J. M. Wefer, R. J. Williams, E. Wieters, G. Woodward, and A. C. Iles. 2019. "Predator Traits Determine Food-Web Architecture across Ecosystems." *Nature Ecology & Evolution* 3:919–927. https://doi.org/10.1038/S41559-019-0899-x.

Brose, U., R. J. Williams, and N. D. Martinez. 2006. "Allometric Scaling Enhances Stability in Complex Food Webs." *Ecology Letters* 9:1228–36.

Brown, J. H. 1995. *Macroecology*. Chicago: University of Chicago Press.

Brown, J. H., J. F. Gillooly, A. P. Allen, V. M. Savage, and G. B. West. 2004. "Toward a Metabolic Theory of Ecology." *Ecology* 85:1771–89.

Brown, J. H., C. A. S. Hall, and R. M. Sibly. 2018. "Equal Fitness Paradigm Explained by a Trade-Off between Generation Time and Energy Production Rate." *Nature Ecology & Evolution* 2:222–68. https://doi.org/10.1038/S41559-017-0430-1.

Brown, J. H., and B. A. Maurer. 1987. "Evolution of Species Assemblages: Effects of Energetic Constraints and Species Dynamics on the Diversification of the North American Avifauna." *American Naturalist* 130:1–17.

———. 1989. "Macroecology: The Division of Food and Space among Species on Continents." *Science* 243: 1445–50.

Brown, J. H., R. M. Sibly, and A. Kodric-Brown. 2012. "Introduction: Metabolism as the Basis for a Theoretical Unification of Ecology." In *Metabolic Ecology: A Scaling Approach*, edited by R. M. Sibly, J. H. Brown, and A. Kodric-Brown, 1–6. Oxford: John Wiley & Sons.

Bull, J. J., and E. L. Charnov. 1988. "How Fundamental Are Fisherian Sex Ratios?" *Oxford Surveys in Evolutionary Biology* 5:96–135.

Burger, J. O., C. Hou, and J. H. Brown. 2019. "Toward a Metabolic Theory of Life History." *Proceedings of the National Academy of Sciences, USA* 116: 26653–61. https://doi.org/10.1073/pnas.1907702116.

Burger, J. O., C. Hou, C. A. S. Hall, and J. H. Brown. 2021. "Universal Rules of Life: Metabolic Rates, Biological Times and the Equal Fitness Paradigm." *Ecology Letters* 24:1262–81. https://doi.org/10.1111/Ele.13715.

Burian, R. M. 1992. "Adaptation: Historical Perspectives." In *Keywords in Evolutionary Biology*, edited by E. F. Keller and E. A. Lloyd, 7–12. Cambridge, MA: Harvard University Press.

Buss, L. W. 1987. *The Evolution of Individuality*. Princeton, NJ: Princeton University Press.

Caldarelli, G. 2007. *Scale-Free Networks: Complex Webs in Nature and Technology*. Oxford: Oxford University Press.

Calder, W. A., III. 1984. *Size, Function, and Life History*. Cambridge, MA: Harvard University Press.

Camazine, S., J.-L. Deneubourg, N. R. Franks, J. Sneyd, G. Theraulaz, and E. Bonabeau. 2001. *Self-Organization in Biological Systems*. Princeton, NJ: Princeton University Press.

Caponi, G. 2016. "Lineages and Systems: A Conceptual Discontinuity in Biological Hierarchies." In *Evolutionary Theory: A Hierarchical Perspective*, edited by N. Eldredge, T. Pievsani, E. Seerelli, and I. Tëmkin, 47–62. Chicago: University of Chicago Press.

Caporael, L. R., J. R. Griesemer, and W. C. Wimsatt, eds. 2014. *Developing Scaffolds in Evolution, Culture, and Cognition*. Cambridge, MA: MIT Press.

Carbone, C., G. M. Mace, S. C. Roberts, and D. W. Macdonald. 1999. "Energetic Constraints on the Diet of Terrestrial Carnivores." *Nature* 402:286–88.

Carbone, C., A. Teacher, and J. M. Rowcliff. 2007. "The Costs of Carnivory." *PLOS Biology* 5:E22. https://doi.org/10.1371/Journal.Pbio.0050022.

Carvalho, A. C., M. C. Sampaio, F. L. Vaandas, and L. B. Klaczko. 1998. "An Experimental Demonstration of Fisher's Principle: Evolution of Sexual Proportion by Natural Selection." *Genetics* 148:719–31. https://doi.org/10.1093/Genetics /148.2.719.

Cartwright, N. 1983. *How the Laws of Physics Lie.* Oxford: Oxford University Press.

———. 1998. "How Theories Relate: Takeovers or Partnerships?" *Philosophia Naturalis* 35: 23–34.

———. 1999. *The Dappled World: A Study of the Boundaries of Science.* Cambridge: Cambridge University Press.

Cassan, A., D. Kubas, J.-P. Beaulieu, M. Dominik, K. Horne, J. Greenhill, J. Wambsganss, J. Menzies, A. Williams, U. G. Jorgensen, A. Udalski, D. P. Bennett, M. D. Albrow, V. Batista, S. Brillant, J. A. R. Caldwell, A. Cole, C. Coutures, K. H. Cook, S. Dieters, D. D. Prester, J. Donatowicz, P. Fouque, K. Hill, N. Kains, S. Kane, J.-B. Marquette, R. Martin, K. R. Pollard, K. C. Sahu, C. Vinter, D. Warren, B. Watson, M. Zub, T. Sumi, M. K. Szymanski, M. Kubiak, R. Poleski, I. Soszynski, K. Ulaczyk, G. Pietrzynski, and L. Wyrzykowski. 2012. "One or More Bound Planets per Milky Way Star from Microlensing Observations." *Nature* 481:167–69.

Cerdá, X., X. Arnan, and J. Retana. 2013. "Is Competition a Significant Hallmark of Ant (Hymenoptera: Formicidae) Ecology?" *Myrmecological News* 18:131–47.

Charnov, E. L. 1993. *Life History Invariants: Some Explorations of Symmetry in Evolutionary Ecology.* Oxford: Oxford University Press.

Charnov, E. L., J. Haskell, and S. K. M. Ernest. 2001. "Density-Dependent Invariance, Dimensionless Life Histories and the Energy-Equivalence Rule." *Evolutionary Ecology Research* 3:117–27.

Charnov, E. L., R. Warne, and M. E. Moses. 2007. "Lifetime Reproductive Effort." *American Naturalist* 170:E129–E142.

Charlesworth, B., M. Nordborg, and D. Charlesworth. 1997. "The Effects of Local Selection, Balanced Polymorphism and Background Selection on Equilibrium Patterns of Genetic Diversity in Subdivided Populations." *Genetic Research, Cambridge* 70:155–74. https://doi.org/10.1017/S0016672397002954.

Chen, J., T. Källman, X. Ma, N. Gyllenstrand, G. Zaina, M. Morgante, J. Bousquet, A. Eckert, J. Wegrzyn, D. Neale, U. Lagerkrantz, and M. Lascoux. 2012. "Disentangling the Roles of History and Local Selection in Shaping Clinical Variation of Allele Frequencies and Gene Expression in Norway Spruce (*Picea abies*)." *Genetics* 191:865–81. https://doi.org/10.1534/Genetics.112.140749.

Cirtwell, A. R., and K. L. Wooton. 2022. "Stable Motifs Delay Species Loss in Simulated Food Webs." *Oikos* 2022:E09436. https://doi.org/10.1111/Oik.09436.

Clarke, B. C. 1979. "Evolution of Genetic Diversity." *Proceedings of the Royal Society of London, Series B, Biological Science* 205:453–74.

Clarke, E. 2014. "Origins of Evolutionary Transitions." *Journal of Biosciences* 39:303–17. https://doi.org/10.1007/S12038-013-9375-Y.

Cleland, C. E. 2011. "Prediction and Explanation in Historical Natural Science." *British Journal for the Philosophy of Science* 62:551–82.

Cockell, C. S. 2020. *Astrobiology: Understanding Life in the Universe*. 2nd ed. Hoboken, NJ: Wiley-Blackwell.

Cohen, J. E. 1971. "Mathematics as Metaphor." *Science* 172:674–75.

Cohen, J. E., F. Briand, and C. M. Newman. 1990. *Community Food Webs: Data and Theory*. Berlin: Springer-Verlag.

Cohen, J. E., T. Jonsson, and S. R. Carpenter. 2003. "Ecological Community Description Using the Food Web, Species Abundance, and Body Size." *Proceedings of the National Academy of Science, USA* 100:1781–91.

Cohen, J. E., S. L. Pimm, P. Yodzis, and J. Saldaña. 1993. "Body Sizes of Animal Predators and Animal Prey in Food Webs." *Journal of Animal Ecology* 62:67–78.

Colyvan, M., J. Damuth, and L. R. Ginzburg. 2019. "The Dawn of Universal Ecology." *Scientist* 33: 20–21.

Colyvan, M., and L. R. Ginzburg. 2003. "Laws of Nature and Laws of Ecology." *Oikos* 101:649–53.

Cordes, C. 2006. "Darwinism in Economics: From Analogy to Continuity." *Journal of Evolutionary Economics* 16:529–41.

Corsi, P. 1978. "The Importance of French Transformist Ideas for the Second Volume of Lyell's *Principles of Geology*." *British Journal for the History of Science* 11:221–44.

———. 2005. "Before Darwin: Transformist Concepts in European Natural History." *Journal of the History of Biology* 38:67–83.

Cotgreave, P. 1993. "The Relationship between Body Size and Population Abundance in Animals." *Trends in Ecology and Evolution* 8:244–48.

———. 1995. "Population Density, Body Mass and Niche Overlap in Australian Birds." *Functional Ecology* 9:285–89.

Cracraft, J. 1982. "A Nonequilibrium Theory for the Rate-Control of Speciation and Extinction and the Origin of Macroevolutionary Patterns." *Systematic Zoology* 31:348–65.

Currie, A. 2019. *Scientific Knowledge & the Deep Past: History Matters*. Cambridge: Cambridge University Press.

Cuzzi, J. N., J. A. Burns, S. Charnoz, R. N. Clark, J. E. Colwell, L. Dones, L. W. Esposito, G. Filacchione, R. G. French, M. M. Hedman, S. Kempf, E. A. Marouf, C. D. Murray, P. D. Nicholson, C. C. Porco, J. Schmidt, M. R. Showalter, L. J. Spilker, J. N. Spitale, R. Srama, M. Sremčević, M. S. Tiscareno, and J. Weiss. 2010. "An Evolving View of Saturn's Rings." *Science* 327:1470–75.

Cziko, G. 1995. *Without Miracles: Universal Selection Theory and the Second Darwinian Revolution*. Cambridge, MA: MIT Press.

Dahlsjö, C. A. L., C. L. Parr, Y. Malhi, P. Meir, H. Rahman, and P. Eggleton. 2015.

"Density-Body Mass Relationships: Inconsistent Intercontinental Patterns among Termite Feeding-Groups." *Acta Oecologica* 63:16–21. https://doi.org/10.1016/J .Actao.2015.01.003.

Dambacher, J. M., H.-K. Luh, H. W. Li, and P. A. Rossignol. 2003. "Qualitative Stability and Ambiguity in Model Ecosystems." *American Naturalist* 161:876–88.

Damuth, J. 1981a. "Home Range, Home Range Overlap, and Species Energy Use among Herbivorous Mammals." *Biological Journal of the Linnean Society* 15: 185–93.

———. 1981b. "Population Density and Body Size in Mammals." *Nature* 290:699–700.

———. 1985. "Selection among 'Species': A Formulation in Terms of Natural Functional Units." *Evolution* 39:1132–46.

———. 1987. "Interspecific Allometry of Population Density in Mammals and Other Animals: The Independence of Body Mass and Population Energy-Use." *Biological Journal of the Linnean Society* 31:193–246.

———. 1988. "Higher Level and Multilevel Selection." Unpublished manuscript.

———. 1993. "Cope's Rule, the Island Rule and the Scaling of Mammalian Population Density." *Nature* 365:748–50.

———. 1998. "Common Rules for Animals and Plants." *Nature* 395:115–16.

———. 2001. "Scaling of Growth: Plants and Animals Are Not So Different." *Proceedings of the National Academy of Science, USA* 98:2113–14.

———. 2007. "A Macroevolutionary Explanation of Energy Equivalence in the Scaling of Body Size and Population Density." *American Naturalist* 169:621–31.

Damuth, J., and I. L. Heisler. 1988. "Alternative Formulations of Multilevel Selection." *Biology and Philosophy* 3:407–30.

Darwin, C. 1859. *On the Origin of Species by Means of Natural Selection, or the Preservation of Favoured Races in the Struggle for Life.* London: John Murray.

Dawkins, R. 1976. *The Selfish Gene.* Oxford: Oxford University Press.

———. 1978. "Replicator Selection and the Extended Phenotype." *Zeitschrift Für Tierpsychologie* 47:61–76.

———. 1983. "Universal Darwinism." In *Evolution from Molecules to Men*, edited by D. S. Bendall, 403–25. Cambridge: Cambridge University Press.

———. 1999. *The Extended Phenotype: The Long Reach of the Gene.* Oxford: Oxford University Press.

Day, E. H., X. Hua, and L. Bromham. 2016. "Is Specialization an Evolutionary Dead End? Testing for Differences in Speciation, Extinction and Trait Transition Rates across Diverse Phylogenies of Specialists and Generalists." *Journal of Evolutionary Biology* 29:1257–67.

De Vladar, H. P., M. Santos, and E. Szathmáry. 2017. "Grand Views of Evolution." *Trends in Ecology and Evolution* 32:324–34.

DeLong, J. P. 2020. "Detecting the Signature of Body Mass Evolution in the Broad-Scale Architecture of Food Webs." *American Naturalist* 196:443–53. https://doi .org/10.1086/710350.

DeLong, J. P., and D. A. Vasseur. 2011. "Mutual Interference Is Common and Mostly Intermediate in Magnitude." *BMC Ecology* 11:1. https://doi.org/10.1186/1472-6785-11-1.

———. 2012. "A Dynamic Explanation of Size-Density Scaling in Carnivores." *Ecology* 93:470–76.

Demetrius, L. 2004. "Caloric Restriction, Metabolic Rate, and Entropy." *Journal of Gerontology* 59A:902–15.

Dennett, D. C. 1995. *Darwin's Dangerous Idea.* New York: Simon and Schuster.

Diamond, J. 1986. "Biology of Birds of Paradise and Bowerbirds." *Annual Review of Ecology and Systematics* 17:17–37.

Dobzhansky, T. 1968. "Adaptedness and Fitness." In *Population Biology and Evolution,* edited by R. C. Lewontin, 109–21. Syracuse, NY: Syracuse University Press.

———. 1970. *Genetics of the Evolutionary Process.* New York: Columbia University Press.

Dodds, W. K. 2009. *Laws, Theories and Patterns in Ecology.* Berkeley: University of California Press.

Doebeli, M., and J. C. Koella. 1995. "Evolution of Simple Population Dynamics." *Proceedings of the Royal Society of London, Series B, Biological Science* 260: 119–25.

Doolittle, W. F. 1981. "Is Nature Really Motherly?" *Coevolution Quarterly* 29:58–62.

———. 2014. "Natural Selection through Survival Alone." *Biology and Philosophy* 29:415–23.

———. 2017. "Darwinizing Gaia." *Journal of Theoretical Biology* 434:11–19.

Doolittle, W. F., and S. A. Inkpen. 2018. "Processes and Patterns of Interaction as Units of Selection: An Introduction to ITSNTS Thinking." *Proceedings of the National Academy of Sciences, USA* 115:4006–14. https://doi.org/10.1073/pnas.1722232115.

Du Toit, J. T., and N. Owen-Smith. 1989. "Body Size, Population Metabolism, and Habitat Specialization among Large African Herbivores." *American Naturalist* 133:736–40.

Dugan, J. E., D. M. Hubbard, and H. M. Page. 1995. "Scaling Population Density to Body Size: Tests in Two Soft-Sediment Intertidal Communities." *Journal of Coastal Research* 11:849–57.

Duncan, R. P., D. M. Forsyth, and J. Hone. 2007. "Testing the Metabolic Theory of Ecology: Allometric Scaling Exponents in Mammals." *Ecology* 88:324–33.

Dykhuizen, D., and D. L. Hartl. 1980. "Selective Neutrality of 6PGD Allozymes in *E. Coli* and the Effects of Genetic Background." *Genetics* 96:801–17.

Eddington, A. S. 1928. *The Nature of the Physical World.* Cambridge: Cambridge University Press.

Edelaar, P., J. Otsuka, and V. J. Luque. 2022. "A Generalised Approach to the Study and Understanding of Adaptive Evolution." *Biological Reviews* 98:352–75. https://doi.org/10.1111/Brv.12910.

Edwards, A. W. F. 1998. "Natural Selection and the Sex Ratio: Fisher's Sources." *American Naturalist* 151:564–69.

Eichholz, J. C. 2014. *Adaptive Capacity: How Organizations Can Thrive in a Changing World.* Greenwich, CT: LID Publishing.

Eldredge, N. 1985. *Unfinished Synthesis: Biological Hierarchies and Modern Evolutionary Thought.* Oxford: Oxford University Press.

——. 1989. *Macroevolutionary Dynamics: Species, Niches, and Adaptive Peaks.* New York: McGraw-Hill.

——. 1999. *The Pattern of Evolution.* New York: W. H. Freeman.

——. 2003. "The Sloshing Bucket: How the Physical Realm Controls Evolution." In *Evolutionary Dynamics: Exploring the Interplay of Selection, Accident, Neutrality, and Function*, edited by J. P. Crutchfield and P. Schuster, 3–32. Oxford: Oxford University Press.

Eldredge, N., and S. N. Salthe. 1984. "Hierarchy and Evolution." In *Oxford Surveys in Evolutionary Biology*, vol. 1., edited by R. Dawkins and M. Ridley, 184–208. Oxford: Oxford University Press.

Eldredge, N., J. N. Thompson, P. M. Brakefield, S. Gavrilets, D. Jablonski, J. B. C. Jackson, R. E. Lenski, B. S. Lieberman, M. A. McPeek, and W. Miller III. 2005. "The Dynamics of Evolutionary Stasis." *Paleobiology* 31 (2, Supplement): 133–45.

Ellner, S., and P. Turchin. 1995. "Chaos in a Noisy World: New Methods and Evidence from Time-Series Analysis." *American Naturalist* 145:343–75.

Elton, C. S. 1927. *Animal Ecology.* London: Sidgwick & Jackson.

Emmerson, M. C., and D. Raffaelli. 2004. "Predator-Prey Body Size, Interaction Strength and Stability of a Real Food Web." *Journal of Animal Ecology* 73: 399–409.

Endler, J. A. 1986. *Natural Selection in the Wild.* Princeton, NJ: Princeton University Press.

Enquist, B. J., J. H. Brown, and G. B. West. 1998. "Allometric Scaling of Plant Energetics and Population Density." *Nature* 395:163–65.

Ernest, S. K. M., B. J. Enquist, J. H. Brown, E. L. Charnov, J. F. Gillooly, V. M. Savage, E. P. White, F. A. Smith, E. A. Hadly, J. P. Haskell, S. K. Lyons, B. A. Maurer, K. J. Niklas, and B. Tiffney. 2003. "Thermodynamic and Metabolic Effects on the Scaling of Production and Population Energy Use." *Ecology Letters* 6:990–95.

Evans, A. R. 2014. "Commentary: Population Density and Body Size in Mammals (1981)." In *Foundations of Macroecology: Classic Papers with Commentaries*, edited by F. Smith, J. Gittleman, and J. Brown, 170–71. Chicago: University of Chicago Press.

Fairbairn, D. J., and J. P. Reeve. 2001. "Natural Selection." In *Evolutionary Ecology: Concepts and Case Studies*, edited by C. W. Fox, D. A. Roff, and D. J. Fairbairn, 29–43. Oxford: Oxford University Press.

Farlow, J. O. 1976. "A Consideration of the Trophic Dynamics of a Late Cretaceous Large-Dinosaur Community (Oldman Formation)." *Ecology* 57:841–57.

Farlow, J. O., D. Coroian, P. J. Currie, J. R. Foster, J. C. Mallon, and F. Therrien. 2023. "'Dragons' on the Landscape: Modeling the Abundance of Large Carnivorous Dinosaurs of the Upper Jurassic Morrison Formation (USA) and the Upper Cretaceous Dinosaur Park Formation (Canada)." *Anatomical Record* 306:1669–96. https://doi.org/10.1002/ar.25024.

Fenchel, T. 1974. "Intrinsic Rate of Natural Increase: The Relationship with Body Size." *Oecologia* 14:317–26.

Fincham, J. R. S. 1972. "Heterozygous Advantage as a Likely General Basis for Enzyme Polymorphisms." *Heredity* 28: 387–91.

Fisher, R. A. 1930. *The Genetical Theory of Natural Selection*. Oxford: Oxford University Press.

Fontaneto, D., C. Q. Tang, U. Obertegger, F. Leasi, and T. G. Barraclough. 2012. "Different Diversification Rates between Sexual and Asexual Organisms." *Evolutionary Biology* 39:262–70. https://doi.org/10.1007/S11692-012-9161-z.

Fortuna, M. A., D. B. Stouffer, J. M. Olesen, P. Jordano, D. Mouillot, B. R. Krasnov, R. Poulin, and J. Bascompte. 2010. "Nestedness versus Modularity in Ecological Networks: Two Sides of the Same Coin?" *Journal of Animal Ecology* 79:811–17. https://doi.org/10.1111/J.1365-2656.2010.01688.x.

Fowler, C. W. 1981. "Density Dependence as Related to Life History Strategy." *Ecology* 62:602–10.

———. 1988. "Population Dynamics as Related to Rate of Increase per Generation." *Evolutionary Ecology* 2:197–204.

Fracchia, J., and R. C. Lewontin. 1999. "Does Culture Evolve?" *History and Theory* 38:52–78. https://doi.org/10.1111/0018-2656.00104.

———. 2005. "The Price of Metaphor." *History and Theory* 44:14–29. https://doi.org/10.1111/J.1468-2303.2005.00305.x.

Frank, S. A. 1995. "George Price's Contributions to Evolutionary Genetics." *Journal of Theoretical Biology* 175:373–88.

———. 1996. "The Design of Natural and Artificial Adaptive Systems." In *Adaptation*, edited by M. R. Rose and G. V. Lauder, 451–505. San Diego, CA: Academic Press.

———. 2012. "Natural Selection III: Selection versus Transmissiion and the Levels of Selection." *Journal of Evolutionary Biology* 25:227–43. https://doi.org/10.1111/J.1420-9101.2011.02431.x.

French, A. R., and T. B. Smith. 2005. "Importance of Body Size in Determining Dominance Hierarchies among Diverse Tropical Frugivores." *Biotropica* 37:96–101.

Fusco, G. 2001. "How Many Processes Are Responsible for Phenotypic Evolution?" *Evolution & Development* 3 (4): 279–86.

Futuyma, D. T., and G. Moreno. 1988. "The Evolution of Ecological Specialization." *Annual Review of Ecology and Systematics* 19:207–33.

Galton, F. 1879. "The Geometric Mean, in Vital and Social Statistics." *Proceedings of the Royal Society of London* 29 (196–99): 365–67. https://doi.org/10.1098/rspl.1879.0060.

Gardner, A. 2008. "Primer: The Price Equation." *Current Biology* 18:R198–R202.

Gardner, A., and J. P. Conlon. 2013. "Cosmological Natural Selection and the Purpose of the Universe." *Complexity* 18:48–56.

Gardner, A., and A. Grafen. 2009. "Capturing the Superorganism: A Formal Theory of Group Adaptation." *Journal of Evolutionary Biology* 22:659–71. https://doi.org/10.1111/J.1420-9101.2008.01681.x.

Gaston, K. J., T. M. Blackburn, and J. H. Lawton. 1993. "Comparing Animals and Automobiles: A Vehicle for Understanding Body Size and Abundance Relationships in Species Assemblages?" *Oikos* 66:172–79.

Getz, W. M. 2006. "The 'Theory of Evolution' Is a Misnomer." *BioScience* 56:96–97.

Giere, R. N. 1999. *Science without Laws*. Chicago: University of Chicago Press.

———. 2008. "Models, Metaphysics, and Methodology." In *Nancy Cartwright's Philosophy of Science*, edited by S. Hartmann, C. Hoefer, and L. Bovens, 123–33. New York: Routledge.

Gillooly, J. F., J. H. Brown, G. B. West, V. M. Savage, and E. L. Charnov. 2001. "Effects of Size and Temperature on Metabolic Rate." *Science* 293:2248–51.

Gingerich, P. D. 1987. "Evolution and the Fossil Record: Patterns, Rates, and Processes." *Canadian Journal of Zoology* 65:1053–60.

Ginzburg, L. R. 1979. "Why Are Heterozygotes Often Superior in Fitness?" *Theoretical Population Biology* 15: 264–67.

———. 1986. "The Theory of Population Dynamics: I, Back to First Principles." *Journal of Theoretical Biology* 122:385–99.

Ginzburg, L. R., O. Burger, and J. Damuth. 2010. "The May Threshold and Life-History Allometry." *Biology Letters* 6:850–53.

Ginzburg, L. R., and M. Colyvan. 2004. *Ecological Orbits: How Planets Move and Populations Grow*. New York: Oxford University Press.

Ginzburg, L. R., and J. Damuth. 2008. "The Space-Lifetime Hypothesis: Viewing Organisms in Four Dimensions, Literally." *American Naturalist* 171: 125–31.

———. 2022. "The Issue Isn't Which Model of Consumer Interference Is Right, but Which One Is Least Wrong." *Frontiers in Ecology and Evolution* 10:860542. https://doi.org/10.3389/fevo.2022.860542.

Ginzburg, L. R., and R. D'Andrea. 2023. "Trophic Levels." In *Encyclopedia of Biodiversity (Fifth Edition)*, edited by S. M. Scheiner. Academic Press. https://www.sciencedirect.com/referencework/9780323984348/encyclopedia-of-biodiversity#book-info.

Gittenberger, E. 1991. "What about Non-Adaptive Radiation?" *Biological Journal of the Linnean Society* 43:263–72.

Glazier, D. S. 2005. "Beyond the '3/4-Power Law': Variation in the Intra- and Interspecific Scaling of Metabolic Rate in Animals." *Biological Reviews* 80:611–62.

———. 2015. "Is Metabolic Rate a Universal 'Pacemaker' for Biological Processes?" *Biological Reviews* 90:377–407.

Goodnight, C. J. 1990a. "Experimental Studies of Community Evolution I: The Response to Selection at the Community Level." *Evolution* 44:1614–24.

———. 1990b. "Experimental Studies of Community Evolution II: The Ecological Basis of the Response to Community Selection." *Evolution* 44:1625–36.

———. 2013. "On Multilevel Selection and Kin Selection: Contextual Analysis Meets Direct Fitness." *Evolution* 67:1539–48.

Goodnight, C. J., J. M. Schwartz, and L. Stevens. 1992. "Contextual Analysis of Models of Group Selection, Soft Selection, Hard Selection, and the Evolution of Altruism." *American Naturalist* 140:743–61.

Gould, S. J. 1966. "Allometry and Size in Ontogeny and Phylogeny." *Biological Reviews* 41:587–640.

———. 1975. "On the Scaling of Tooth Size in Mammals." *American Zoologist* 15: 351–62.

———. 1988. "Trends as Changes in Variance: A New Slant on Progress and Directionality in Evolution." *Journal of Paleontology* 62:319–29.

———. 1997a. "Darwinian Fundamentalism." *New York Review of Books* 44 (10): 34–37.

———. 1997b. "Cope's Rule as Psychological Artefact." *Nature* 385:199–200.

———. 1998. "Gulliver's Further Travels: The Necessity and Difficulty of a Hierarchical Theory of Selection." *Philosophical Transactions of the Royal Society of London, Series B, Biological Sciences* 353:307–14.

———. 2002. *The Structure of Evolutionary Theory*. Cambridge, MA: Harvard University Press.

Gould, S. J., and R. Lewontin. 1978. "The Spandrels of San Marco and the Panglossian Paradigm: A Critique of the Adaptationist Programme." *Proceedings of the Royal Society of London* 205:581–98.

Grantham, T. A. 1995. "Hierarchical Approaches to Macroevolution: Recent Work on Species Selection and the 'Effect Hypothesis.'" *Annual Review of Ecology and Systematics* 26:301–21.

Griesemer, J. 2000. "The Units of Evolutionary Transition." *Selection* 1:67–80.

———. 2013. "Formalization and the Meaning of 'Theory' in the Inexact Biological Sciences." *Biological Theory* 7:298–310.

Griesemer, J., and A. Shavit. 2023. "Scaffolding Individuality: Coordination, Cooperation, Collaboration and Community." *Philosophical Transactions of the Royal Society of London, Series B, Biological Sciences* 378:20210398. https://doi.org/10.1098/rstb.2021.0398.

Griesemer, J. R. 1984. "Presentations and the Status of Theories." In *PSA: Proceedings of the Biennial Meeting of the Philosophy of Science Association, 1984*, vol. 1, edited by P. D. Asquith, 102–14. East Lansing, MI: Philosophy of Science Association.

———. 1990a. "Material Models in Biology." In *PSA 1990: Proceedings of the Biennial Meeting of the Philosophy of Science Association*, vol. 2, *Symposia and Invited Papers.*, edited by A. Fine, M. Forbes and L. Wessels, 79–93. Chicago: University of Chicago Press.

———. 1990b. "Modeling in the Museum: On the Role of Remnant Models in the Work of Joseph Grinnell." *Biology and Philosophy* 5:3–36.

———. 2004. "Three-Dimensional Models in Philosophical Perspective." In *Models: The Third Dimension of Science*, edited by S. de Chadarevian and N. Hopwood, 433–42. Stanford, CA: Stanford University Press.

Grilli, J., T. Rogers, and S. Allesina. 2016. "Modularity and Stability in Ecological Communities." *Nature Communications* 7:12031. https://doi.org/10.1038/ncomms12031.

Guimarães, P. R., P. Jordano, and J. N. Thompson. 2011. "Evolution and Coevolution in Mutualistic Networks." *Ecology Letters* 14:877–85. https://doi.org/10.1111/j.1461-0248.2011.01649.x.

Guo, G., G. Barabás, G. Takimoto, D. Bearup, W. F. Fagan, D. Chen, and J. Liao. 2023. "Towards a Mechanistic Understanding of Variation in Aquatic Food Chain Length." *Ecology Letters* 26:1926–39. https://doi.org/10.1111/ele.14305.

Hallett, J. G. 1991. "The Structure and Stability of Small Mammal Faunas." *Oecologia* 88:383–93.

Hamilton, W. D. 1964. "The Genetical Evolution of Social Behavior." *Journal of Theoretical Biology* 7:1–16.

———. 1967. "Extraordinary Sex Ratios." *Science* 156:477–88.

———. 1975. "Innate Social Aptitudes of Man: An Approach from Evolutionary Genetics." In *Biosocial anthropology*, edited by R. Fox, 133–53. London: Malaby Press.

Hamilton, W. D., and T. M. Lenton. 1998. "Spora and Gaia: How Microbes Fly with Their Clouds." *Ethology Ecology & Evolution* 10:1–16. https://doi.org/10.1080/08927014.1998.9522867.

Hanski, I. A., and M. E. Gilpin, eds. 1997. *Metapopulation Biology: Ecology, Genetics, and Evolution*. San Diego, CA: Academic Press.

Harman, D. 1956. "Ageing: A Theory Based on Free Radical and Radiation Chemistry." *Journal of Gerontology* 11:298–300.

Haskell, J. P., M. E. Ritchie, and H. Olff. 2002. "Fractal Geometry Predicts Varying Body Size Scaling Relationships for Mammal and Bird Home Ranges." *Nature* 418:527–30.

Hassell, M. P., and G. C. Varley. 1969. "New Inductive Population Model for Insect Parasites and Its Bearing on Biological Control." *Nature* 223:1133–37.

Hatton, I. A., A. P. Dobson, D. Storch, E. D. Galbraith, and M. Loreau. 2019. "Linking Scaling Laws across Eukaryotes." *Proceedings of the National Academy of Sciences, USA* 116:21616–22.

Healy, K., T. Guillerme, S. Finlay, A. Kane, S. B. A. Kelly, D. McClean, D. J. Kelly, I. Donohue, A. L. Jackson, and N. Cooper. 2014. "Ecology and Mode-of-Life Explain Lifespan Variation in Birds and Mammals." *Proceedings of the Royal Society of London, Series B, Biological Science* 281:20140298. https://doi.org/http://dx.doi.org/10.1098/rspb.2014.0298.

Hechinger, R. F., K. D. Lafferty, A. P. Dobson, J. H. Brown, and A. M. Kuris. 2011. "A Common Scaling Rule for Abundance, Energetics, and Production of Parasitic and Free-Living Species." *Science* 333:445–48.

Heisler, I. L., and J. Damuth. 1987. "A Method for Analyzing Selection in Hierarchically Structured Populations." *American Naturalist* 130:582–602.

Hemmingsen, A. M. 1950. "The Relation of Standard (Basal) Energy Metabolism to Total Fresh Weight of Living Organisms." *Reports of the Steno Memorial Hospital and the Nordisk Insulinlaboratorium* 4:7–38.

———. 1960. "Energy Metabolism as Related to Body Size and Respiratory Surfaces, and Its Evolution." *Reports of the Steno Memorial Hospital and the Nordisk Insulinlaboratorium* 9:1–110.

Hendry, A. P. 2017. *Eco-Evolutionary Dynamics*. Princeton, NJ: Princeton University Press.

Hodgson, G. M. 2002. "Darwinism in Economics." *Journal of Evolutionary Economics* 12:259–81.

Hoefer, C. 2008. "For Fundamentalism." In *Nancy Cartwright's Philosophy of Science*, edited by S. Hartmann, C. Hoefer, and L. Bovens, 306–21. New York: Routledge.

Holland, J. H. 1992. *Adaptation in Natural and Artificial Systems: An Introductory Analysis with Applications to Biology, Control, and Artificial Intelligence*. Cambridge, MA: MIT Press.

———. 1995. *Hidden Order: How Adaptation Builds Complexity*. Reading, MA: Addison-Wesley.

Holt, R. D. 1977. "Predation, Apparent Competition, and the Structure of Prey Communities." *Theoretical Population Biology* 12:197–29.

———. 1997. "Community Modules." In *Multitrophic Interactions in Terrestrial Ecosystems*, edited by A. C. Gange and V. K. Brown, 333–50. Oxford: Blackwell.

Holt, R. D. and M. B. Bonsall. 2017. "Apparent Competition." *Annual Review of Ecology, Evolution and Systematics* 48:447–71. https://doi.org/10.1146/annurev-ecolsys-110316-022628.

Hubbell, S. P. 2001. *The Unified Neutral Theory of Biodiversity and Biogeography*. Princeton, NJ: Princeton University Press.

Hubbell, S. P., and J. K. Lake. 2004. "The Neutral Theory of Biodiversity and Biogeography, and Beyond." In *Macroecology: Concepts and Consequences*, edited by T. M. Blackburn and K. J. Gaston, 45–63. Cambridge: Cambridge University Press.

Hui, C. and D. M. Richardson. 2019. "How to Invade an Ecological Network." *Trends in Ecology and Evolution* 34:121–31. https://doi.org/10.1016/j.tree.2018.11.003.

Hull, D. L. 1980. "Individuality and Selection." *Annual Review of Ecology and Systematics* 11:311–32.

———. 1988. *Science as a Process: An Evolutionary Account of the Social and Conceptual Development of Science*. Chicago: University of Chicago Press.

Hull, D. L. and M. Ruse. 1998. Introduction to *The Philosophy of Biology*, edited by D. L. Hull and M. Ruse, 1–2. Oxford: Oxford University Press.

Humphreys, W. F. 1979. "Production and Respiration in Animal Populations." *Journal of Animal Ecology* 48:427–53.

———. 1981. "Towards a Simple Index Based on Live Weight and Biomass to Predict Assimilation in Natural Populations." *Journal of Animal Ecology* 50:543–61.

Huxley, J. 1942. *Evolution: The Modern Synthesis*. New York: Harper.

Huxley, J. S. 1932. *Problems of Relative Growth*. New York: Dial Press.

Huxley, J. S., and G. Teissier. 1936. "Terminology of Relative Growth." *Nature* 137: 780–81.

Jablonski, D. 1986. "Larval Ecology and Macroevolution in Marine Invertebrates." *Bulletin of Marine Science* 39:565–87.

———. 1987. "Heritability at the Species Level: Analysis of Geographic Ranges of Cretaceous Mollusks." *Science* 238:360–63.

———. 1997. "Body-Size Evolution in Cretaceous Molluscs and the Status of Cope's Rule." *Nature* 385:250–52. https://doi.org/10.1038/385199a0.

———. 2008. "Species Selection: Theory and Data." *Annual Review of Ecology, Evolution and Systematics* 39:501–24.

———. 2017. "Approaches to Macroevolution: 2, Sorting of Variation, Some Overarching Issues, and General Conclusions." *Evolutionary Biology* 44:451–75. https://doi.org/10.1007/s11692-017-9434-7.

Jablonski, D., and G. Hunt. 2006. "Larval Ecology, Geographic Range, and Species Survivorship in Cretaceous Mollusks: Organismic versus Species-Level Explanations." *American Naturalist* 168:556–64.

Jablonski, D., and R. A. Lutz. 1983. "Larval Ecology of Marine Benthic Invertebrates: Paleobiological Implications." *Biological Reviews* 58:21–89.

Jablonski, D., and J. J. Sepkoski. 1996. "Paleobiology, Community Ecology, and Scales of Ecological Pattern." *Ecology* 77:1367–78.

Jeler, C. 2020. "Explanatory Goals and Explanatory Means in Multilevel Selection Theory." *History and Philosophy of the Life Sciences* 42:36. https://doi.org/10.1007/s40656-020-00333-y.

Jensen, C. X. J., and L. R. Ginzburg. 2005. "Paradoxes or Theoretical Failures? The Jury Is Still Out." *Ecological Modelling* 188:3–14.

Jetz, W., C. Carbone, J. Fulford, and J. H. Brown. 2004. "The Scaling of Animal Space Use." *Science* 306:266–68.

Johnson, S., V. Domínguez-García, L. Donetti, and M. A. Muñoz. 2014. "Trophic Coherence Determines Food-Web Stability." *Proceedings of the National Academy of Sciences, USA* 111:17923–28. https://doi.org/10.1073/pnas.1409077111.

Jonsson, T., and B. Ebenman. 1998. "Effects of Predator-Prey Body Size Ratios on the Stability of Food Chains." *Journal of Theoretical Biology* 193:407–17.

Jonsson, T., J. E. Cohen, and S. R. Carpenter. 2005. "Food Webs, Body Size, and

Species Abundance in Ecological Community Description." In *Advances in Ecological Research*, Vol. 36, edited by H. Caswell, 1–84. Amsterdam: Elsevier.

Jordano, P. 1987. "Patterns of Mutualistic Interactions in Pollination and Seed Dispersal: Connectance, Dependence Asymmetries, and Coevolution." *American Naturalist* 129:657–77.

Juanes, F. 1986. "Population Density and Body Size in Birds." *American Naturalist* 128:921–29.

Kauffman, S. A. 1993. *The Origins of Order: Self-Organization and Selection in Evolution*. Oxford: Oxford University Press.

Kawecki, T., and D. Ebert. 2004. "Conceptual Issues in Local Adaptation." *Ecology Letters* 7:1225–41. https://doi.org/10.1111/j.1461-0248.2004.00684.x.

Keddy, P. A., and B. Shipley. 1989. "Competitive Hierarchies in Herbaceous Plant Communities." *Oikos* 54:234–41.

Keddy, P. A. 2001. *Competition*. 2nd ed. Dordrecht: Kluwer Academic.

Keller, E. F. 2005. "Ecosystems, Organisms, and Machines." *BioScience* 55:1069–74.

———. 2007. "The Disappearance of Function from 'Self-Organizing Systems.'" In *Systems Biology: Philosophical Foundations*, edited by F. C. Boogerd, F. J. Bruggeman, J.-H. S. Hofmeyr, and H. V. Westerhoff, 303–17. Amsterdam: Elsevier.

Kendall, M., and A. Stuart. 1979. *The Advanced Theory of Statistics*. 4th ed. 3 vols. London: Griffin.

Kerkhoff, A. J., and B. J. Enquist. 2009. "Multiplicative by Nature: Why Logarithmic Transformation Is Necessary in Allometry." *Journal of Theoretical Biology* 257:519–21. https://doi.org/10.1016/j.jtbi.2008.12.026.

Kingsolver, J. G., H. H. Hoekstra, J. M. Hoekstra, D. Berrigan, S. N. Vignieri, S. E. Hill, A. Hoang, P. Gilbert, and P. Beerli. 2001. "The Strength of Phenotypic Selection in Natural Populations." *American Naturalist* 157:245–61.

Kinlock, N. L. 2019. "A Meta-Analysis of Plant Interaction Networks Reveals Competitive Hierarchies as Well as Facilitation and Intransitivity." *American Naturalist* 194:640–53.

Kirchner, J. W. 2002. "The Gaia Hypothesis: Fact, Theory, and Wishful Thinking." *Climatic Change* 52:391–408.

Kirkpatrick, M. 1982. "Sexual Selection and the Evolution of Female Choice." *Evolution* 36:1–12.

Kleiber, M. 1975. *The Fire of Life*. New York: Krieger.

Knouft, J. H. 2002. "Regional Analysis of Body Size and Population Density in Stream Fish Assemblages: Testing Predictions of the Energetic Equivalence Rule." *Canadian Journal of Zoology* 59: 1350–60.

Kondoh, M., S. Kato, and Y. Sakato. 2010. "Food Webs Are Built up with Nested Subwebs." *Ecology* 91:3123–30.

Krakauer, D. C., ed. 2019. *Worlds Hidden in Plain Sight: The Evolving Idea of Complexity at the Santa Fe Institute*. Santa Fe, NM: Santa Fe Institute Press.

Kuhry, B., and L. F. Marcus. 1977. "Bivariate Linear Models in Allometry." *Systematic Zoology* 26:201–9.

Kuris, A. M., R. F. Hechinger, J. C. Shaw, K. L. Whitney, L. Aguirre-Macedo, C. A. Boch, A. P. Dobson, E. J. Dunham, B. L. Fredensborg, T. C. Huspeni, J. Lorda, L. Mababa, F. T. Mancini, A. B. Mora, M. Pickering, N. L. Talhouk, M. E. Torchin, and K. D. Lafferty. 2008. "Ecosystem Energetic Implications of Parasite and Free-Living Biomass in Three Estuaries." *Nature* 454:515–18.

Kutschera, U. 2003. "A Comparative Analysis of the Darwin-Wallace Papers and the Development of the Concept of Natural Selection." *Theory in Biosciences* 122:343–59.

Lack, D. 1954. *The Natural Regulation of Animal Numbers.* Oxford: Clarendon Press.

Laird, R. A., and B. S. Schamp. 2006. "Competitive Intransitivity Promotes Species Coexistence." *American Naturalist* 168:182–93.

Lande, R. 1980. "Sexual Dimorphism, Sexual Selection, and Adaptation in Polygenic Characters." *Evolution* 34:292–305.

———. 1981. "Models of Speciation by Sexual Selection on Polygenic Traits." *Proceedings of the National Academy of Sciences, USA* 78:3721–25.

Lande, R., and S. J. Arnold. 1983. "The Measurement of Selection on Correlated Characters." *Evolution* 37:1210–36.

Latour, B. 1987. *Science in Action: How to Follow Scientists and Engineers through Society.* Cambridge, MA: Harvard University Press.

Laughlin, R. B. 2005. *A Different Universe: Reinventing Physics from the Bottom Down.* Cambridge, MA: Basic Books.

Laughlin, R. B., and D. Pines. 2000. "The Theory of Everything." *Proceedings of the National Academy of Sciences, USA* 97: 28–31.

Legendre, P., and L. Legendre. 1998. *Numerical Ecology.* 2nd ed. Amsterdam: Elsevier.

Lehtonen, J. 2018. "The Price Equation, Gradient Dynamics, and Continuous Trait Game Theory." *American Naturalist* 191:146–53.

Lenton, T. 2016. *Earth System Science: A Very Short Introduction.* Oxford: Oxford University Press.

Lenton, T. M. 1998. "Gaia and Natural Selection." *Nature* 394:439–47.

Lenton, T. M., T. A. Kohler, P. A. Marquet, R. A. Boye, M. Crucifix, D. M. Wilkinson, and M. Scheffer. 2021. "Survival of the Systems." *Trends in Ecology and Evolution* 36:333–44. https://doi.org/10.1016/j.tree.2020.12.003.

Lerner, I. M. 1954. *Genetic Homeostasis.* New York: Wiley.

Levin, S. A. 1998. "Ecosystems and the Biosphere as Complex Adaptive Systems." *Ecosystems* 1:431–36.

Levin, S. A., and C. P. Goodyear. 1980. "Analysis of an Age-Structured Fishery Model." *Journal of Mathematical Biology* 9:245–74.

Levins, R. 1968. *Evolution in Changing Environments: Some Theoretical Explorations.* Princeton, NJ: Princeton University Press.

———. 1970. Complex Systems. In *Organization, Stability and Process: Toward a Theoretical Biology*, vol. 3, edited by C. H. Waddington, 73–88. New Brunswick, NJ: Transaction Publishers.

Levinton, J. S. 2001. *Genetics, Paleontology, and Macroevolution*. Cambridge: Cambridge University Press.

Lewontin, R. C. 1970. "The Units of Selection." *Annual Review of Ecology and Systematics* 1:1–18.

———. 1978. "Adaptation." *Scientific American* 239:156–69.

———. 1980. "Adaptation." In *Encyclopedia Einaudi*, reprinted in E. Sober, ed., *Conceptual Issues in Evolutionary Biology*. Cambridge, MA: MIT Press; Milan: Einaudi, 1984.

Lewontin, R. C., L. R. Ginzburg, and S. D. Tuljapurkar. 1978. "Heterosis as an Explanation for Large Amounts of Genic Polymorphism." *Genetics* 88:149–69.

Lewontin, R. C., and R. Levins. 2007. *Biology under the Influence: Dialectical Essays on Ecology, Agriculture, and Health*. New York: Monthly Review Press.

Li, C. C. 1967. "Fundamental Theorem of Natural Selection." *Nature* 214:505–6.

Lindstedt, S. L., and W. A. Calder III. 1981. "Body Size, Physiological Time, and Longevity of Homeothermic Mammals." *Quarterly Review of Biology* 56:1–16.

Lloyd, E. A. 1988. *The Structure and Confirmation of Evolutionary Theory*. New York: Greenwood Press.

———. 1994. "Rx: Distinguish Group Selection from Group Adaptation." *Behavioral and Brain Sciences* 17:628–29.

———. 2001. "Units and Levels of Selection: Anatomy of the Units of Selection Debates." In *Thinking about Evolution: Historical, Philosophical, and Political Perspectives*, edited by R. S. Singh, C. B. Krimbas, D. B. Paul, and J. Beatty, 267–291. Cambridge: Cambridge University Press.

———. 2008. "An Open Letter to Elliott Sober and David Sloan Wilson, Regarding Their Book, *Unto Others, the Evolution and Psychology of Unselfish Behavior*." In *Science, Evolution and Politics*, edited by E. A. Lloyd, 95–105. Cambridge: Cambridge University Press.

———. 2012. "Units and Levels of Selection." In *The Stanford Encyclopedia of Philosophy*, edited by E. N. Zalta. Metaphysics Research Lab, Stanford University. http://plato.stanford.edu/archives/win2012/entries/selection-units/.

———. 2015. "Evolutionary Theory, Structure Of." In *International Encyclopedia of the Social and Behavioral Sciences*, 2nd ed., vol. 8, edited by J. D. Wright, 469–75. Burlington: Elsevier.

Lloyd, E. A., and S. J. Gould. 1993. "Species Selection on Variability." *Proceedings of the National Academy of Science, USA* 90:595–99.

Loehle, C. 1988. "Philosophical Tools: Potential Contributions to Ecology." *Oikos* 51:97–104.

Loeuille, N., and M. Loreau. 2005. "Evolutionary Emergence of Size-Structured Food Webs." *Proceedings of the National Academy of Sciences, USA* 102:5761–66. https://doi/10.1073?pnas.0408424102.

Logofet, D. O. 1993. *Matrices and Graphs: Stability Problems in Mathematical Ecology*. Boca Raton, FL: CRC Press.

———. 2005. "Stronger-Than-Lyapunov Notions of Matrix Stability, or How 'Flowers' Help Solve Problems in Mathematical Ecology." *Linear Algebra and its Applications* 398:75–100.

Lotka, A. J. 1925. *Elements of Physical Biology*. Baltimore, MD: Williams and Wilkins.

Love, A. C., and W. C. Wimsatt, eds. 2019. *Beyond the Meme: Development and Structure in Cultural Evolution*. Mineapolis: University of Minnesota Press.

Lovelock, J. E. 1972. "Gaia as Seen through the Atmosphere." *Atmospheric Environment* 6:579–80.

———. 1979. *Gaia: A New Look at Life on Earth*. Oxford: Oxford University Press.

Lovelock, J. E., and L. Margulis. 1974. "Atmospheric Homeostasis by and for the Biosphere: The Gaia Hypothesis." *Tellus* 26:2–10. https://doi.org/10.3402/tellusa.v26i1-2.9731.

Lucretius. 2001. *On the Nature of Things*. Translated by M. F. Smith. Indianapolis: Hackett.

Luque, V. J. 2017. "One Equation to Rule Them All: A Philosophical Analysis of the Price Equation." *Biology and Philosophy* 32:97–125.

Lyell, C. 1830. *Principles of Geology, Being an Attempt to Explain the Former Changes of the Earth's Surface, by Reference to Causes Now in Operation*. Vol. 1. London: John Murray.

Lynch, M. 2007a. "The Frailty of Adaptive Hypotheses for the Origins of Organismal Complexity." *Proceedings of the National Academy of Science, USA* 104: 8597–604.

———. 2007b. "The Evolution of Genetic Networks by Non-Adaptive Processes." *Nature Reviews Genetics* 8:803–13.

Lynch, M., and W. G. Hill. 1986. "Phenotypic Evolution by Neutral Mutation." *Evolution* 40:915–35.

Magnani, L., N. J. Nersessian, and P. Thagard, eds. 1999. *Model-Based Reasoning in Scientific Discovery*. New York: Springer.

Maiorana, V. C. 1990. "Evolutionary Strategies and Body Size in a Guild of Mammals." In *Body Size in Mammalian Paleobiology: Estimation and Biological Implications*, edited by J. Damuth and B. J. MacFadden, 69–102. New York: Cambridge University Press.

Marbà, N., C. M. Duarte, and S. Agustí. 2007. "Allometric Scaling of Plant Life History." *Proceedings of the National Academy of Science, USA* 104:15777–80.

Marcy, A. E., T. Guillerme, E. Sherratt, K. C. Rowe, M. J. Phillips, and V. Weisbecker. 2020. "Australian Rodents Reveal Conserved Cranial Evolutionary Allometry across 10 Million Years of Murid Evolution." *American Naturalist* 196:755–68. https://doi.org/10.1086/711398.

Margulis, L. 1981. "Gaia Lives, Has Blurred Boundaries." *CoEvolution Quarterly* 29:63–65.

Marquet, P. A., S. A. Navarrette, and J. C. Castilla. 1990. "Scaling Population Density to Body Size in Rocky Intertidal Communities." *Science* 250:1125–27.

———. 1995. "Body Size, Population Density, and the Energetic Equivalence Rule." *Journal of Animal Ecology* 64:325–32.

Marshall, C. R. 2017. "Five Palaeobiological Laws Needed to Understand the Evolution of the Living Biota." *Nature Ecology & Evolution* 1:0615. https://doi.org/10.1038/s41559-017-0165.

Marshall, C. R., T. V. Latorre, C. J. Wilson, T. M. Frank, K. M. Magoulick, J. B. Zimmt, and A. W. Poust. 2021. "Absolute Abundance and Preservation Rate of *Tyrannosaurus rex*." *Science* 372:284–87. https://doi.org/10.1126/science.abc8300.

Marshall, J. A. R. 2011. "Group Selection and Kin Selection: Formally Equivalent Approaches." *Trends in Ecology and Evolution* 26:325–32.

May, R. M. 1973a. *Stability and Complexity in Model Ecosystems*. Princeton, NJ: Princeton University Press.

———. 1973b. "Qualitative Stability in Model Ecosystems." *Ecology* 54:638–41.

———. 1974. "Biological Populations with Nonoverlapping Generations: Stable Points, Stable Cycles, and Chaos." *Science* 186:645–47.

May, R. M., and G. F. Oster. 1976. "Bifurcations and Dynamic Complexity in Simple Ecological Models." *American Naturalist* 110:573–99.

Maynard, D. S., C. A. Serván, and S. Allesina. 2018. "Network Spandrels Reflect Ecological Assembly." *Ecology Letters* 21:324–34. https://doi.org/10.1111/ele.12912.

Maynard Smith, J. 1964. "Group Selection and Kin Selection." *Nature* 201:1145–47.

Maynard Smith, J., and E. Szathmáry. 1995. *The Major Transitions in Evolution*. New York: W. H. Freeman.

———. 1996. "On the Likelihood of Habitable Worlds." *Nature* 384:107.

Mayr, E. 1963. *Animal Species and Evolution*. Cambridge, MA: Harvard University Press.

———. 1972. "Sexual Selection and Natural Selection." In *Sexual Selection and the Descent of Man*, edited by B. G. Campbell, 87–104. Chicago: Aldine.

———. 1983. "How to Carry Out the Adaptationist Program?" *American Naturalist* 121:324–34.

McGowran, B., and Q. Li. 2000. "Evolutionary Palaeoecology of Cainozoic Foraminifera: Tethys, Indo-Pacific, Southern Australasia." *Historical Biology* 15:3–27.

McLaren, I. A., ed. 1971. *Natural Regulation of Animal Populations*. New York: Atherton Press.

McNab, B. K. 1963. "Bioenergetics and the Determination of Home Range Size." *American Naturalist* 97:133–40.

McShea, D. W., and R. Brandon. 2010. *Biology's First Law: The Tendency for Diversity and Complexity to Increase in Evolutionary Systems*. Chicago: University of Chicago Press.

Medeiros, L. P., K. Boege, E. del-Val, A. Zaldivar-Riverón, and S. Saavedra. 2021. "Observed Ecological Communities Are Formed by Species Combinations That Are among the Most Likely to Persist under Changing Environments." *American Naturalist* 197:E17–E29. https://doi.org/10.1086/711663.

Medeiros, L. P., G. Garcia, J. N. Thompson, and P. R. Guimarães Jr. 2018. "The Geographic Mosaic of Coevolution in Mutualistic Networks." *Proceedings of the National Academy of Sciences, USA* 115:12017–22. https://doi.org/10.1073/pnas.1809088115.

Meehan, T. D. 2006. "Energy Use and Animal Abundance in Litter and Soil Communities." *Ecology* 87:1650–58.

Meehan, T. D., P. K. Drumm, R. S. Farrar, K. Oral, K. E. Lanier, E. A. Pennington, L. A. Pennington, I. T. Stafurik, D. V. Valore, and A. D. Wylie. 2006. "Energetic Equivalence in a Soil Arthropod Community from an Aspen-Conifer Forest." *Pedobiologia* 50:307–12.

Michod, R. E. 1986. "On Fitness and Adaptedness and Their Role in Evolutionary Explanation." *Journal of the History of Biology* 19:289–302.

———. 2005. "On the Transfer of Fitness from the Cell to the Multicellular Organism." *Biology and Philosophy* 20:967–87. https://doi.org/10.1007/s10539-005-9018-2.

Mikkelson, G. M. 2003. "Ecological Kinds and Ecological Laws." *Philosophy of Science* 70:1390–1400.

Milo, R., S. Shen-Orr, S. Itzkovitz, N. Kashtan, D. Chklovskii, and U. Alon. 2002. "Network Motifs: Simple Building Blocks of Complex Networks." *Science* 298: 824–27.

Mitchell, S. D. 2002. "*Ceteris paribus*—an Inadequate Representation for Biological Contingency." *Erkenntnis* 57:329–50.

Mitchell-Olds, T., and R. G. Shaw. 1987. "Regression Analysis of Natural Selection: Statistical Inference and Biological Interpretation." *Evolution* 41:1149–61.

Mittelbach, G. G., and B. J. McGill. 2019. *Community Ecology.* 2nd ed. New York: Oxford University Press.

Mohr, C. O. 1940. "Comparative Populations of Game, Fur and Other Mammals." *American Midland Naturalist* 24:581–84.

Molles, M. C., Jr. 2019. *Ecology: Concepts and Applications.* 8th ed. New York: McGraw-Hill.

Mora, B. B., D. Gravel, L. J. Gilarranz, T. Poisot, and D. B. Stouffer. 2018. "Identifying a Common Backbone of Interactions Underlying Food Webs from Different Ecosystems." *Nature Communications* 9:2603. https://doi.org/10.1038/s41467-018-05056-0.

Morin, A., and P. Dumont. 1994. "A Simple Model to Estimate Growth Rate of Lotic Insect Larvae and Its Value for Estimating Population and Community Production." *Journal of the North American Benthological Society* 13:357–67.

Morrison, M., and M. S. Morgan. 1999. "Models as Mediating Agents." In *Models as Mediators: Perspectives on Natural and Social Science*, edited by M. S. Morgan and M. Morrison, 10–37. Cambridge: Cambridge University Press.

Morse, D. R., J. H. Lawton, M. M. Dodson, and M. H. Williamson. 1985. "Fractal Dimension of Vegetation and the Distribution of Arthropod Body Lengths." *Nature* 314:731–33.

Moya-Larano, J., J. Rowntree, and G. Woodward, eds. 2014. *Eco-Evolutionary Dynamics*. London: Elsevier.

Nagy, K. A. 2005. "Field Metabolic Rate and Body Size." *Journal of Experimental Biology* 208:1621–25.

Nagy, K. A., I. A. Girard, and T. K. Brown. 1999. "Energetics of Free-Ranging Mammals, Reptiles, and Birds." *Annual Review of Nutrition* 19:247–77.

Nanay, B. 2011. "Replication without Replicators." *Synthese* 179:455–77. https://doi.org/10.1007/s11229-009-9702-x.

National Academy of Sciences. 2008. *Science, Evolution, and Creationism*. Washington, DC: National Academy Press.

Nee, S., A. F. Read, J. J. D. Greenwood, and P. H. Harvey. 1991. "The Relationship between Abundance and Body Size in British Birds." *Nature* 351:312–13.

Nelson, R. R., and S. G. Winter. 1982. *An Evolutionary Theory of Economic Change*. Cambridge, MA: Harvard University Press.

Nersessian, N. J. 1999. "Model-Based Reasoning in Conceptual Change." In *Model-Based Reasoning in Scientific Discovery*, edited by L. Magnani, N. J. Nersessian and P. Thagard, 5–22. New York: Springer.

Neutel, A.-M., J. A. P. Heesterbeek, and P. C. De Ruiter. 2002. "Stability in Real Food Webs: Weak Links in Long Loops." *Science* 296:1120–23.

Neutel, A.-M., and M. A. S. Thorne. 2018. "Symmetry, Asymmetry, and Beyond: The Crucial Role of Interaction Strength in the Complexity-Stability Debate." In *Adaptive Food Webs: Stability and Transitions of Real and Model Ecosystems*, edited by J. C. Moore, P. C. de Ruiter, K. S. McCann, and V. Wolters, 31–44, Cambridge: Cambridge University Press.

Niklas, K. J., and B. J. Enquist. 2001. "Invariant Scaling Relationships for Interspecific Plant Biomass Production Rates and Body Size." *Proceedings of the National Academy of Science, USA* 98:2922–27.

Nitecki, M. H., and D. V. Nitecki, eds. 1992. *History and Evolution*. Albany: State University of New York Press.

Novak, M., and D. B. Stouffer. 2021. "Systematic Bias in Studies of Consumer Functional Responses." *Ecology Letters* 24:580–93. https://doi.org/10.1111/ele.13660.

Novak, M., J. D. Yeakel, A. E. Noble, D. F. Doak, M. Emmerson, J. A. Estes, U. Jacob, M. T. Tinker, and J. T. Wooton. 2016. "Characterizing Species Interactions to Understand Press Perturbations: What Is the Community Matrix?" *Annual Review of Ecology, Evolution and Systematics* 47:409–32.

Odum, E. P. 1968. "Energy Flow in Ecosystems: A Historical Review." *American Zoologist* 8:11–18.

Okasha, S. 2006. *Evolution and the Levels of Selection*. Oxford: Oxford University Press.

———. 2016. "The Relation between Kin and Multilevel Selection: An Approach Using Causal Graphs." *British Journal for the Philosophy of Science* 67:435–70.

Okasha, S., and C. Paternotte. 2012. "Group Adaptation, Formal Darwinism and Contextual Analysis." *Journal of Evolutionary Biology* 25:1127–39.

Olson, E. C. 1952. "The Evolution of a Permian Vertebrate Chronofauna." *Evolution* 6:181–96.

Orr, H. A. 1996. "Dennett's Dangerous Idea." *Evolution* 50:467–72.

Orzack, S. H., and P. Forber. 2017. "Adaptationism." In *The Stanford Encyclopedia of Philosophy*, edited by E. N. Zalta. Metaphysics Research Lab, Stanford University. https://plato.stanford.edu/archives/spr2017/entries/adaptationism.

O'Sullivan, J. D., R. J. Knell, and A. G. Rossberg. 2019. "Metacommunity-Scale Biodiversity Regulation and the Self-Organised Emergence of Macroecological Patterns." *Ecology Letters* 22:1428–38.

Otto, S. P., and A. Rosales. 2020. "Theory in Service to Narratives in Evolution and Ecology." *American Naturalist* 195:290–99. https://doi.org/10.1086/705991.

Paine, R. T. 1980. "Food Webs: Linkage, Interaction Strength and Community Infrastructure." *Journal of Animal Ecology* 49:667–85. https://doi.org/10.2307/4220.

Papale, F. 2021. "Evolution by Means of Natural Selection without Reproduction: Revamping Lewontin's Account." *Synthese* 198:10429–55. https://doi.org/10.1007/s11229-020-02729-6.

Pearl, J. 2018. *The Book of Why: The New Science of Cause and Effect*. New York: Basic Books.

Pearl, J., M. Glymour, and N. P. Jewell. 2016. *Causal Inference in Statistics: A Primer*. West Sussex, UK: Wiley.

Pearl, R. 1928. *The Rate of Living: Being an Account of Some Experimental Studies on the Biology of Life Duration*. New York: Alfred A. Knopf.

Persson, L. 1985. "Assymetrical Competition: Are Larger Animals Competitively Superior?" *American Naturalist* 126:261–66.

Peters, R. H. 1983. *The Ecological Implications of Body Size*. Cambridge: Cambridge University Press.

Peterson, R. O. 2013. "It's a Wonderful Gift." *Science* 339:142–43.

Pigliucci, M. 2013. "On the Different Ways of 'Doing Theory' in Biology." *Biological Theory* 7:287–97.

Pimm, S. L. 1982. *Food Webs*. London: Chapman and Hall.

———. 1991. *The Balance of Nature? Ecological Issues in the Conservation of Species and Communities*. Chicago: University of Chicago Press.

Pimm, S. L., and J. H. Lawton. 1977. "Number of Trophic Levels in Ecological Communities." *Nature* 268:329–31.

Piras, P., D. Silvestro, F. Carotenuto, S. Castiglione, A. Kotsakis, L. Maiorino, M. Melchionna, A. Mondanaro, G. Sansalone, C. Serio, A. V. Vero, and P. Raia. 2018. "Evolution of the Sabertooth Mandible: A Deadly Ecomorphological Specialization." *Palaeogeography, Palaeoclimatology, Palaeoecology* 496:166–74. https://doi.org/10.1016/j.palaeo.2018.01.034.

Polis, G. A., C. A. Myers, and R. D. Holt. 1989. "The Ecology and Evolution of Intraguild Predation: Potential Competitors That Eat Each Other." *Annual Review of Ecology and Systematics* 20:297–330.

Pool, R. 1989. "Ecologists Flirt with Chaos." *Science* 243:310–13.

Popper, K. R. 1972. *Objective Knowledge: An Evolutionary Approach.* Oxford: Clarendon Press.

Post, D. M., and E. P. Palkovacs. 2009. "Eco-Evolutionary Feedbacks in Community and Ecosystem Ecology: Interactions between the Ecological Theatre and the Evolutionary Play." *Proceedings of the Royal Society of London B: Biological Sciences* 364:1629–40.

Price, C. A., J. S. Weitz, V. M. Savage, J. Stegen, A. Clarke, D. A. Coomes, P. S. Dodds, R. S. Etienne, A. J. Kerkhoff, K. McCulloh, K. J. Niklas, H. Olff, and N. G. Swenson. 2012. "Testing the Metabolic Theory of Ecology." *Ecology Letters* 15:1465–74.

Price, G. R. 1970. "Selection and Covariance." *Nature* 227:520–21.

———. 1972. "Extension of Covariance Selection Mathematics." *Annals of Human Genetics* 35:485–90.

———. 1995. "The Nature of Selection." *Journal of Theoretical Biology* 175:389–96.

Proulx, S. R., D. E. L. Promislow, and P. C. Phillips. 2005. "Network Thinking in Ecology and Evolution." *Trends in Ecology and Evolution* 20:345–53.

Queller, D. C. 2017. "Fundamental Theorems of Evolution." *American Naturalist* 189:345–53. https://doi.org/10.1086/690937.

Quirk, J., and R. Ruppert. 1965. "Qualitative Economics and the Stability of Equilibria." *The Review of Economic Studies* 32:311–26.

Raatikainen, P. 2022. "Gödel's Incompleteness Theorems." In *The Stanford Encyclopedia of Philosophy*, edited by E. N. Zalta. Metaphysics Research Lab, Stanford University. https://plato.stanford.edu/archives/spr2022/entries/goedel -incompleteness/.

Rabosky, D. L. 2010. "Primary Controls on Species Richness in Higher Taxa." *Systematic Biology* 59:634–45. https://doi.org/10.1093/sysbio/syq060.

———. 2013. "Diversity-Dependence, Ecological Speciation, and the Role of Competition in Macroevolution." *Annual Review of Ecology, Evolution and Systematics* 44:481–502. https://doi.org/10.1146/annurev-ecolsys-110512-135800.

Rabosky, D. L., and E. E. Goldberg. 2015. "Model Inadequacy and Mistaken Inferences of Trait-Dependent Speciation." *Systematic Biology* 64:340–55.

Rabosky, D. L., and D. R. Matute. 2013. "Macroevolutionary Speciation Rates Are Decoupled from the Evolution of Intrinsic Reproductive Isolation in *Drosophila* and Birds." *Proceedings of the National Academy of Sciences, USA* 110: 15354–59.

Rabosky, D. L., and A. R. McCune. 2010. "Reinventing Species Selection with Molecular Phylogenies." *Trends in Ecology and Evolution* 25:68–74. https://doi.org /10.1016/j.tree.2009.07.002.

Radzvilavicius, A. L., and N. W. Blackstone. 2018. "The Evolution of Individuality Revisited." *Biological Reviews* 93:1620–33. https://doi.org/10.1111/brv.12412.

Rankin, B. D., J. W. Fox, C. R. Barrón-Ortiz, A. E. Chew, P. A. Holroyd, J. A. Ludtke, X. Yang, and J. M. Theodor. 2018. "The Extended Price Equation Quantifies

Species Selection on Mammalian Body Size across the Palaeocene/Eocene Thermal Maximum." *Proceedings of the Royal Society of London B: Biological Sciences* 282:20151097. https://doi.org/10.1098/rspb.2015.1097.

Raup, D. M., and J. J. Sepkoski Jr. 1984. "Periodicity of Extinctions in the Geologic Past." *Proceedings of the National Academy of Science, USA* 81:801–5.

Reeve, H. K., and P. W. Sherman. 1993. "Adaptation and the Goals of Evolutionary Research." *Quarterly Review of Biology* 68:1–32.

Rice, S. H. 2004. *Evolutionary Theory: Mathematical and Conceptual Foundations.* Sunderland, MA: Sinauer.

Ricker, W. E. 1954. "Stock and Recruitment." *Journal of the Fisheries Research Board of Canada* 11:559–623.

Ricklefs, R. E. 2004. "A Comprehensive Framework for Global Patterns in Diversity." *Ecology Letters* 7:1–15.

———. 2015. "Intrinsic Dynamics of the Regional Community." *Ecology Letters* 18: 497–503.

Ricklefs, R. E., and D. Schluter, eds. 1993. *Species Diversity in Ecological Communities: Historical and Geographical Perspectives.* Chicago: University of Chicago Press.

Roberts, A. 1974. "The Stability of a Feasible Random Ecosystem." *Nature* 251: 607–8.

Robertson, A. 1966. "A Mathematical Model of the Culling Process in Dairy Cattle." *Animal Production* 8:95–108.

Rogers, T. L., B. J. Johnson, and S. B. Munch. 2022. "Chaos Is Not Rare in Natural Ecosystems." *Nature Ecology & Evolution* 6:1105–11. https://doi.org/10.1038/s41559-022-01787-y.

Rose, M. R., and G. V. Lauder. 1996. "Concepts and Theories of Adaptation." In *Adaptation*, edited by M. R. Rose and G. V. Lauder, 9. San Diego, CA: Academic Press.

Rosenblum, E. B., B. A. J. Sarver, J. W. Brown, S. Des Roches, K. M. Hardwick, T. D. Hether, J. M. Eastman, M. W. Pennell, and L. J. Harmon. 2012. "Goldilocks Meets Santa Rosalia: An Ephemeral Speciation Model Explains Patterns of Diversification across Time Scales." *Evolutionary Biology* 39:255–61.

Rosenzweig, M. L. 1971. "Paradox of Enrichment: Destabilization of Exploitation Ecosystems in Ecological Time." *Science* 171:385–87.

———. 1973. "Evolution of the Predator Isocline." *Evolution* 27:84–94.

———. 1995. *Species Diversity in Space and Time.* Cambridge: Cambridge University Press.

Rosenzweig, M. L., and R. H. MacArthur. 1968. "Graphical Representation and Stability Conditions of Predator-Prey Interactions." *American Naturalist* 97: 209–23.

Rossberg, A. G. 2013. *Food Webs and Biodiversity: Foundations, Models, Data.* Oxford: John Wiley & Sons.

Rossberg, A. G., R. Ishi, T. Amemiya, and K. Itoh. 2008. "The Top-Down Mechanism for Body-Mass-Abundance Scaling." *Ecology* 89:567–80.

Roughgarden, J. 1983. "The Theory of Coevolution." In *Coevolution*, edited by D. J. Futuyma and M. Slatkin, 33–64. Sunderland, MA: Sinauer.

Roux, S. 2005. "Empedocles to Darwin." In *Greek Research in Australia: Proceedings of the Biennial International Conference of Greek Studies, Flinders University April 2003*, edited by E. Close, T. M. and G. Frazis, 1–16. Adelaide: Flinders University Department of Languages–Modern Greek.

Rubner, M. 1883. "Ueber den Einfluss der Körpergrösse auf Stoff- und Kraftswechsel." *Zeitschrift für Biologie* 19:535–62.

———. 1908. *Das Problem der Lebensdauer und seine Beziehungen zu Wachstum und Ernärung*. Munich: R. Oldenbourg.

Ruse, M. 1970. "Are There Laws in Biology?" *Australasian Journal of Philosophy* 48:234–46.

———. 2013. *The Gaia Hypothesis : Science on a Pagan Planet*. Chicago: University of Chicago Press.

Russo, S. E., S. K. Robinson, and J. Terborgh. 2003. "Size-Abundance Relationships in an Amazonian Bird Community: Implications for the Energetic Equivalence Rule." *American Naturalist* 161:267–83.

Sacher, G. A. 1959. "Relation of Lifespan to Brain Weight and Body Weight in Mammals." In *The Lifespan of Animals: Colloquia on Ageing*, vol. 5, edited by G. E. W. Wolstenholme and M. O'Connor, 115–32. Boston: Little, Brown.

Sanford, E., M. S. Roth, G. C. Johns, J. P. Wares, and G. N. Somero. 2003. "Local Selection and Latitudinal Variation in a Marine Predator-Prey Interaction." *Science* 300:1135–37.

Savage, V. M., J. F. Gillooly, W. H. Woodruff, G. B. West, A. P. Allen, B. J. Enquist, and J. H. Brown. 2004. "The Predominance of Quarter-Power Scaling in Biology." *Functional Ecology* 18:257–82.

Scheiner, S. M. 2010. "Towards a Conceptual Framework for Biology." *Quarterly Review of Biology* 85:293–318.

Schoener, T. W. 1983. "Field Experiments on Interspecific Competition." *American Naturalist* 122:240–85.

Selman, C., J. D. Blount, D. H. Nussey, and R. Speakman. 2012. "Oxidative Damage, Ageing, and Life-History Evolution: Where Now?" *Trends in Ecology and Evolution* 27:570–77.

Sepkoski, D. 2016. "'Replaying Life's Tape': Simulations, Metaphors, and Historicity in Stephen Jay Gould's View of Life." *Studies in History and Philosophy of Biological and Biomedical Sciences* 58:73–81. https://doi.org/10.1016/j.shpsc.2015.12.009.

Sepkoski, J. J., Jr. 1978. "A Kinetic Model of Phanerozoic Taxonomic Diversity: I, Analysis of Marine Orders." *Paleobiology* 4:223–51.

———. 1979. "A Kinetic Model of Phanerozoic Taxonomic Diversity: II, Phanerozoic Families and Multiple Equilibria." *Paleobiology* 5:222–51.

———. 1984. "A Kinetic Model of Phanerozoic Taxonomic Diversity: III, Post-Paleozoic Families and Mass Extinctions." *Paleobiology* 10:246–67.

Sewall, B. J., A. L. Freestone, J. E. Hawes, and E. Andriamanarina. 2013. "Size-Energy Relationships in Ecological Communities." *PLoS ONE* 8:e6857.

Shelton, D. E., and R. E. Michod. 2014a. "Group Selection and Group Adaptation during a Major Evolutionary Transition: Insights from the Evolution of Multicellularity in the Volvocine Algae." *Biological Theory* 9:452–69. https://doi.org /10.1007/s13752-014-0159-x.

———. 2014b. "Levels of Selection and the Formal Darwinism Project." *Biology and Philosophy* 29:217–24.

Shipley, B. 1993. "A Null Model for Competitive Hierarchies in Competition Matrices." *Ecology* 74:1693–99.

Shurin, J. B., P. Amarasekare, J. M. Chase, R. D. Holt, M. F. Hoopes, and M. A. Leibold. 2004. "Alternative Stable States and Regional Community Structure." *Journal of Theoretical Biology* 227:359–68.

Sibly, R. M., D. Barker, M. C. Denham, J. Hone, and M. Pagel. 2005. "On the Regulation of Populations of Mammals, Birds, Fish, and Insects." *Science* 309:607–10.

Silva, M., and J. A. Downing. 1995. "The Allometric Scaling of Density and Body Mass: A Nonlinear Relationship for Terrestrial Mammals." *American Naturalist* 145:704–27.

Simon, B. 2014. "Continuous-Time Models of Group Selection, and the Dynamical Insufficiency of Kin Selection Models." *Journal of Theoretical Biology* 349:22–31. https://doi.org/10.1016/j.jtbi.2014.01.030.

Simon, B., J. A. Fletcher, and M. Doebeli. 2012. "Towards a General Theory of Group Selection." *Evolution* 67:1561–72. https://doi.org/10.1111/j.1558-5646.2012 .01835.x.

Simon, H. A. 1996. *The Sciences of the Artificial.* 3rd ed. Cambridge, MA: MIT Press.

Simpson, C. 2010. "Species Selection and Driven Mechanisms Jointly Generate a Large-Scale Morphological Trend in Monobathrid Crinoids." *Paleobiology* 36: 481–96.

———. 2013. "Species Selection and the Macroevolution of Coral Coloniality and Photosymbiosis." *Evolution* 67:1627–21.

Simpson, G. G. 1944. *Tempo and Mode in Evolution.* New York: Columbia University Press.

———. 1953. *The Major Features of Evolution.* New York: Columbia University Press.

Skalski, G. T., and J. F. Gilliam. 2001. "Functional Responses with Predator Interference: Viable Alternatives to the Holling Type II Model." *Ecology* 82:3083–92.

Smith, F. A., S. K. Lyons, S. K. M. Ernest, and J. H. Brown. 2008. "Macroecology: More Than the Division of Food and Space among Species on Continents." *Progress in Physical Geography* 32:115–38.

Smith, R. J. 1984. "Allometric Scaling in Comparative Biology: Problems of Concept and Method." *American Journal of Physiology* 256:R152–R160.

Smolin, L. 1997. *The Life of the Cosmos.* Oxford: Oxford University Press.

Sober, E. 1984. *The Nature of Selection: Evolutionary Theory in Philosophical Focus.* Cambridge, MA: MIT Press.

———. 1988. *Reconstructing the Past.* Cambridge, MA: MIT Press.

———. 1993. *Philosphy of Biology.* Boulder, CO: Westview Press.

———. 1997. "Two Outbreaks of Lawlessness in Recent Philosophy of Biology." *Philosophy of Science* 64:S458–S467.

———. 2000. *Philosophy of Biology.* 2nd ed. New York: Routledge.

———. 2011. *Did Darwin Write the Origin Backwards? Philosophical Essays on Darwin's Theory.* Amherst, NY: Prometheus Books.

Sober, E., and D. S. Wilson. 1998. *Unto Others: The Evolution and Psychology of Unselfish Behavior.* Cambridge, MA: Harvard University Press.

———. 2011. "*Adaptation and Natural Selection* Revisited." *Journal of Evolutionary Biology* 24: 462–68.

Sohal, R. S., and R. G. Allen. 1990. "Oxidative Stress as a Causal Factor in Differentiation and Aging: A Unifying Hypothesis." *Experimental Gerontology* 25:499–522.

Solé, R. V., and J. Bascompte. 2006. *Self-Organization in Complex Ecosystems.* Princeton, NJ: Princeton University Press.

Soliveres, S., F. T. Maestre, W. Ulrich, P. Manning, S. Boch, M. A. Bowker, D. Prati, M. Delgado-Baquerizo, J. L. Quero, I. Schöning, A. Gallardo, W. Weisser, J. Müller, S. A. Socher, M. García-Gómez, V. Ochoa, E. Schulze, M. Fischer, and E. Allan. 2015. "Intransitive Competition Is Widespread in Plant Communities and Maintains Their Species Richness." *Ecology Letters* 18:790–98.

Speakman, J. R. 2005. "Body Size, Energy Metabolism and Lifespan." *Journal of Experimental Biology* 208:1717–30.

Speakman, J. R., C. Selman, J. S. McLaten, and E. J. Harper. 2002. "Living Fast, Dying When? The Link between Ageing and Energetics." *Journal of Nutrition* 132:1583S–97S.

Staley, J. T., and G. H. Orians. 2000. "Evolution and the Biosphere." In *Earth System Science: From Biogeochemical Cycles to Global Change,* edited by M. C. Jacobson, R. J. Charlson, H. Rodhe, and G. H. Orians, 29–67. London: Academic Press.

Staniczenko, P. P. A., J. C. Kopp, and S. Allesina. 2013. "The Ghost of Nestedness in Ecological Networks." *Nature Communications* 4:1391. https://doi.org/10.1038/ncomms2422.

Stanley, S. M. 1975. "A Theory of Evolution above the Species Level." *Proceedings of the National Academy of Sciences, USA* 72:646–50.

———. 1979. *Macroevolution, Pattern and Process.* San Francisco: W. H. Freeman.

Stevens, C. F. 2009. "Darwin and Huxley Revisited: The Origin of Allometry." *Journal of Biology* 8 (Art. 14): 1–7.

Stouffer, D. B., E. L. Rezende, and L. A. N. Amaral. 2011. "The Role of Body Mass in Diet Contiguity and Food-Web Structure." *Journal of Animal Ecology* 80:632–39. https://doi.org/10.1111/j.1365-2656.2011.01812.x.

Sugihara, G. 2017. *Niche Hierarchy: Structure, Organization, and Assembly in Natural Systems*. Plantation, FL: J. Ross Publishing.

Suweis, S., F. Simini, J. R. Banavar, and A. Maritan. 2013. "Emergence of Structural and Dynamical Properties of Ecological Mutualistic Networks." *Nature* 500:449–52. https://doi.org/10.1038/nature12438.

Szathmáry, E. 2015. "Toward Major Evolutionary Transitions Theory 2.0." *Proceedings of the National Academy of Sciences, USA* 112:10104–11.

Taylor, P. D., and S. A. Frank. 1996. "How to Make a Kin Selection Model." *Journal of Theoretical Biology* 180:27–37.

Taylor, P. J. 1987. "Historical versus Selectionist Explanations in Evolutionary Biology." *Cladistics* 3:1–13.

Teller, P. 2004. "The Law-Idealization." *Philosophy of Science* 71:730–41.

Thomas, W. R., M. J. Pomerantz, and M. E. Gilpin. 1980. "Chaos, Asymmetric Growth and Group Selection for Dynamical Stability." *Ecology* 61:1312–20.

Thompson, D. W. 1917. *On Growth and Form*. Cambridge: Cambridge University Press.

Thompson, J. N. 2005. *The Geographic Mosaic of Coevolution*. Chicago: University of Chicago Press.

———. 2006. "Mutualistic Webs of Species." *Science* 312:372–73. https://doi.org/10.1126/science.1126904.

———. 2013. *Relentless Evolution*. Chicago: University of Chicago Press.

Thompson, J. N., S. L. Nuismer, and R. Gomulkiewicz. 2002. "Coevolution and Maladaptation." *Integrative and Comparative Biology* 42:381–87.

Thompson, P. 1989. *The Structure of Biological Theories*. Albany: State University of New York Press.

Thompson, R. M., M. Hemberg, B. M. Starzomski, and J. B. Shurin. 2007. "Trophic Levels and Trophic Tangles: The Prevalence of Omnivory in Real Food Webs." *Ecology* 88:612–17.

Toman, J., and J. Flegr. 2017. "Stability-Based Sorting: The Forgotten Process behind (Not Only) Biological Evolution." *Journal of Theoretical Biology* 435:29–41.

Tomiya, S. 2013. "Body Size and Extinction Risk in Terrestrial Mammals above the Species Level." *American Naturalist* 182:E196–E214. https://doi.org/10.1086/673489.

Toulmin, S. 1972. *Human Understanding: The Collective Use and Evolution of Concepts*. Princeton, NJ: Princeton University Press.

Tran, J. K. 2011. *Adaptation by Group Selection: Elimination Acts on All Levels*. Stony Brook, NY: Stony Brook University.

Traulsen, A. 2009. "Mathematics of Kin- and Group-Selection: Formally Equivalent?" *Evolution* 64:316–23.

Tregonning, K., and A. Roberts. 1978. "Ecosystem-Like Behavior of a Random-Interaction Model I." *Bulletin of Mathematical Biology* 40:513–24.

———. 1979. "Complex Systems Which Evolve towards Homeostasis." *Nature* 281:563–64.

Tuljapurkar, S., C. Boe, and K. W. Wachter. 1994. "Nonlinear Feedback Dynamics in Fisheries: Analysis of the Deriso-Schnute Model." *Canadian Journal of Fisheries and Aquatic Sciences* 51:1462–73.

Turchin, P. 2001. "Does Population Ecology Have General Laws?" *Oikos* 94:17–26.

Turelli, M., and L. R. Ginzburg. 1983. "Should Individual Fitness Increase with Heterozygosity?" *Genetics* 104:191–209.

Tyutyunov, Y., D. Sen, and M. Banerjee. 2024. "Does Mutual Interference Stabilize Prey–Predator Model with Bazykin–Crowley–Martin Trophic Function?" *Mathematical Biosciences* 372: 109201.

Ulanowicz, R. E. 2001. "Information Theory in Ecology." *Computers & Chemistry* 25:393–99.

Ulanowicz, R. E., R. D. Holt, and M. Barfield. 2014. "Limits on Ecosystem Complexity: Insights from Ecological Network Analysis." *Ecology Letters* 17:127–36.

Ulrich, W., S. Soliveres, W. Kryszewski, F. T. Maestre, and N. J. Gotelli. 2014. "Matrix Models for Quantifying Competitive Intransitivity from Species Abundance Data." *Oikos* 123:1057–70.

Upham, N., J. A. Esselstyn, and W. Jetz. 2020. "Ecological Causes of Uneven Speciation and Species Richness in Mammals." Preprint (version 4). Accessed August23, 2020. https://doi.org/10.1101/504803.

Van Valen, L. M. 1960. "Nonadaptive Aspects of Evolution." *American Naturalist* 94:305–8.

——. 1973. "A New Evolutionary Law." *Evolutionary Theory* 1:1–30.

——. 1976. "Energy and Evolution." *Evolutionary Theory* 1:179–229.

——. 1980a. "Evolution as a Zero-Sum Game for Energy." *Evolutionary Theory* 4:289–300.

——. 1980b. "Patch Selection, Benefactors, and a Revitalization of Ecology." *Evolutionary Theory* 4: 231–33.

——. 1989. "Three Paradigms of Evolution." *Evolutionary Theory* 9:1–17.

——. 1991. "Biotal Evolution: A Manifesto." *Evolutionary Theory* 10:1–13.

Van Valkenburgh, B. 1991. "Iterative Evolution of Hypercarnivory in Canids (Mammalia: Carnivora): Evolutionary Interactions among Sympatric Predators." *Paleobiology* 17:340–62.

——. 1999. "Major Patterns in the History of Carnivorous Mammals." *Annual Review of Earth and Planetary Sciences* 27:463–93.

——. 2007. "*Déjà Vu*: The Evolution of Feeding Morphologies in the Carnivora." *Integrative and Comparative Biology* 47:147–63. https://doi.org/10.1093/icb/icm016.

Van Valkenburgh, B., and I. Jenkins. 2002. "Evolutionary Patterns in the History of Permo-Triassic and Cenozoic Synapsid Predators." In *The Fossil Record of Predation*, edited by M. Kowalewski and P. H. Kelley, 267–88. The Paleontological Society Papers 8. Washington, DC: The Paleontological Society.

Van Valkenburgh, B., X. Wang, and J. Damuth. 2004. "Cope's Rule, Hypercarnivory, and Extinction in North American Canids." *Science* 306:101–4.

Velasco, J. A., and J. N. Pinto-Ledezma. 2022. "Mapping Species Diversification Metrics in Macroecology: Prospects and Challenges." *Frontiers in Ecology and Evolution* 10:951271. https://doi.org/10.3389/fevo.2022.951271.

Vellend, M. 2010. "Conceptual Synthesis in Community Ecology." *Quarterly Review of Biology* 85:183–206.

———. 2016. *The Theory of Ecological Communities*. Princeton, NJ: Princeton University Press.

Vermeij, G. J., and E. G. Leigh Jr. 2011. "Natural and Human Economies Compared." *Ecosphere* 2:Art. 39. https://doi.org/10.1890/ES11-00004.1.

Villmoare, B. 2023. *The Evolution of Everything: The Patterns and Causes of Big History*. Cambridge: Cambridge University Press.

Volterra, V. 1926. "Variazioni e fluttuazioni del numero d'individui in specie animali conviventi." *Memorie Accademia dei Lincei* 2:31–113.

Vrba, E. D. 1980. "Evolution, Species and Fossils: How Does Life Evolve?" *South African Journal of Science* 76:61–83.

Vrba, E. S. 1984. "What Is Species Selection?" *Systematic Zoology* 33:318–28.

Vrba, E. S., and N. Eldredge. 1984. "Individuals, Hierarchies and Processes: Towards a More Complete Evolutionary Theory." *Paleobiology* 10:146–71.

Vrba, E. S., and S. J. Gould. 1986. "The Hierarchical Expansion of Sorting and Selection: Sorting and Selection Cannot Be Equated." *Paleobiology* 12:217–28.

Wade, M. J. 1978. "A Critical Review of the Models of Group Selection." *Quarterly Review of Biology* 53:101–14.

———. 1979. "The Evolution of Social Interactions by Family Selection." *American Naturalist* 113:399–417.

———. 1980. "Kin Selection and Its Components." *Science* 210:665–67.

———. 1985. "Soft Selection, Hard Selection, Kin Selection, and Group Selection." *American Naturalist* 125:61–73.

———. 2016. *Adaptation in Metapopulations: How Interaction Changes Evolution*. Chicago: University of Chicago Press.

Wade, M. J., and S. Kalisz. 1990. "The Causes of Natural Selection." *Evolution* 44:1947–55.

Wade, M. J., D. S. Wilson, C. Goodnight, D. Taylor, Y. Bar-Yam, M. A. M. de Aguiar, B. Stacey, J. Werfel, G. A. Hoelzer, E. D. Brodie III, P. Fields, F. Breden, T. A. Linksvayer, J. A. Fletcher, P. J. Richerson, J. D. Bever, V. D. J. D., and P. Zee. 2010. "Multilevel and Kin Selection in a Connected World." *Nature* 463:E8–E9. https://doi.org/10.1038/nature08809.

Wagner, G. P. 2014. *Homology, Genes, and Evolutionary Innovation*. Princeton, NJ: Princeton University Press.

Wallace, A. R. 1858. "On the Tendency of Varieties to Depart Indefinitely from the Original Type." *Biological Journal of the Linnean Society* 3:53–62.

Ward, P. D., and D. Brownlee. 2000. *Rare Earth: Why Complex Life Is Uncommon in the Universe*. New York: Copernicus Books.

Warren, P. H., and J. H. Lawton. 1987. "Invertebrate Predator-Prey Body Size Relationships: An Explanation for Upper Triangular Food Webs and Patterns in
Food Web Structure?" *Oecologia* 74:231–35.

Webb, S. D., and N. D. Opdyke. 1995. "Global Climatic Influence on Cenozoic Land
Mammal Faunas." In *Effects of Past Global Change on Life*, edited by J. Ken-
nett and S. Stanley, 184–208. Washington, DC: National Academy Press.

Weitz, J. S., and S. A. Levin. 2006. "Size and Scaling of Predator-Prey Dynamics."
Ecology Letters 9:548–57. https://doi.org/10.1111/j.1461-0248.2006.00900.x.

West, G. B., J. H. Brown, and B. J. Enquist. 1997. "A General Model for the Origin
of Allometric Scaling Laws in Biology." *Science* 276:122–26.

———. 1999. "The Fourth Dimension of Life: Fractal Geometry and Allometric
Scaling of Organisms." *Science* 284:1677–79.

West-Eberhard, M. J. 1992. "Adaptation: Current Usages." In *Keywords in Evolutionary Biology*, edited by E. F. Keller and E. A. Lloyd, 13–18. Cambridge, MA:
Harvard University Press.

White, E. P., S. K. M. Ernest, A. J. Kerkhoff, and B. J. Enquist. 2007. "Relationships
between Body Size and Abundance in Ecology." *Trends in Ecology and Evolution* 22:323–30.

Whittaker, R. H. 1975. *Communities and Ecosystems*. New York: Macmillan.

Whyte, L. L. 1960. "Developmental Selection of Mutations." *Science* 132:954.

Wickens, A. P. 2001. "Ageing and the Free Radical Theory." *Respiration Physiology*
128:379–91.

Wiens, J. A. 1966. "On Group Selection and Wynne-Edwards' Hypothesis." *American Scientist* 54:273–87.

Wild, G., A. Gardner, and S. A. West. 2010. "Reply to Wade, et al. 2010." *Nature*
463:E9. https://doi.org/10.1038/nature08810.

Wilkinson, D. M. 1999. "Is Gaia Really Conventional Ecology?" *Oikos* 84:533–36.

Williams, G. C. 1966. *Adaptation and Natural Selection*. Princeton, NJ: Princeton
University Press.

———. 1992a. "Gaia, Nature Worship and Biocentric Fallacies." *Quarterly Review
of Biology* 67:479–86.

———. 1992b. *Natural Selection: Domains, Levels, and Challenges*. New York: Oxford University Press.

Williamson, M. 1972. *The Analysis of Biological Populations*. London: Edward Arnold.

Wilson, D. S. 1980. *The Natural Selection of Populations and Communities*. Menlo
Park, CA: Benjamin/Cummings.

———. 1983. "The Group Selection Controversy: History and Current Status." *Annual Review of Ecology and Systematics* 14:159–87.

———. 1997. "Biological Communities as Functionally Organized Units." *Ecology*
78:2018–24.

Wilson, D. S., and W. Swenson. 2003. "Community Genetics and Community Selection." *Ecology* 84:586–88.

Wimsatt, W. C. 1980. "Reductionistic Research Strategies and Their Biases in the

Units of Selection Controversy." In *Scientific Discovery: Case Studies*, edited by T. Nickles, 213–59. Dordrecht: D. Reidel.

Wing, S. L., H.-D. Sues, R. Potts, W. A. DiMichele, and A. K. Behrensmeyer. 1992. "Evolutionary Paleoecology." In *Terrestrial Ecosystems through Time: Evolutionary Paleoecology of Terrestrial Plants and Animals*, edited by A. K. Behrensmeyer, J. D. Damuth, W. A. DiMichele, R. Potts, H.-D. Sues, and S. L. Wing, 1–11. Chicago: University of Chicago Press.

Winn, J. N., and D. C. Fabrycky. 2015. "The Occurrence and Architecture of Exoplanetary Systems." *Annual Review of Astronomy and Astrophysics* 53:409–47.

Winther, R. G. 2021. "The Structure of Scientific Theories." In *The Stanford Encyclopedia of Philosophy*, edited by N. Zalta. Metaphysics Research Lab, Stanford University. https://plato.stanford.edu/archives/spr2021/entries/structure-scientific-theories/.

Winther, R. G., M. J. Wade, and C. C. Dimond. 2013. "Pluralism in Evolutionary Controversies: Styles and Averaging Strategies in Hierarchical Selection Theories." *Biology and Philosophy* 28:957–79.

Wong, M. L., C. E. Cleland, D. Arend Jr., S. Bartlett, H. J. Cleaves II., H. Demarest, A. Prabhu, J. L. Lunine, and R. M. Hazen. 2023. "On the Roles of Function and Selection in Evolving Systems." *Proceedings of the National Academy of Sciences, USA* 120:1–11. https://doi.org/10.1073/pnas.2310223120.

Woodward, G., B. Ebenman, M. Emmerson, J. M. Montoya, J. M. Olesen, A. Valido, and P. H. Warren. 2005. "Body Size in Ecological Networks." *Trends in Ecology and Evolution* 20:402–9.

Woodward, J. 2002. "There Is No Such Thing as a *Ceteris Paribus* Law." *Erkenntnis* 57:303–28.

———. 2019. "Scientific Explanation." In *The Stanford Encyclopedia of Philosophy*, edited by E. N. Zalta. Metaphysics Research Lab, Stanford University. https://plato.stanford.edu/archives/win2019/entries/scientific-explanation.

Wooton, J. T., and M. Emmerson. 2005. "Measurement of Interaction Strength in Nature." *Annual Review of Ecology, Evolution and Systematics* 36:419–44.

Wynne-Edwards, V. C. 1962. *Animal Dispersion in Relation to Social Behaviour*. London: Oliver and Boyd.

———. 1965. "Self-Regulating Systems in Populations of Animals." *Science* 147:1543–48.

———. 1986. *Evolution through Group Selection*. Oxford: Blackwell Scientific Publications.

Yeakel, J. D., C. P. Kempes, and S. Redner. 2018. "Dynamics of Starvation and Recovery Predict Extinction Risk and Both Damuth's Law and Cope's Rule." *Nature Communications* 9:Art. 65. https://doi.org/10.1038/s41467-018-02822-y.

Yodzis, P., and S. Innes. 1992. "Body Size and Consumer-Resource Dynamics." *American Naturalist* 139:1151–75.

Zirkle, C. 1941. "Natural Selection before the 'Origin of Species.'" *Proceedings of the American Philosophical Society* 84:71–123.

Index

Page numbers followed by *f* indicate figures.